The Children's Blizzard

ALSO BY DAVID LASKIN

Artists in Their Gardens (coauthor)
Partisans
Rains All the Time
Braving the Elements
A Common Life

THE
Children's
Blizzard

David Laskin

HARPER PERENNIAL

NEW YORK • LONDON • TORONTO • SYDNEY

HARPER ● PERENNIAL

A hardcover edition of this book was published in 2004 by HarperCollins Publishers.

P.S.™ is a trademark of HarperCollins Publishers.

HarperCollins books may be purchased for educational, business, or sales promotional use. For information please write: Special Markets Department, HarperCollins Publishers, 10 East 53rd Street, New York, NY 10022.

First Harper Perennial edition published in 2005.

Map by Megan Ernst
Designed by Jaime Putorti

Library of Congress Cataloging-in-Publication Data is available upon request.

ISBN-10: 0-06-052076-0 (pbk.)
ISBN-13: 978-0-06-052076-2 (pbk.)

05 06 07 08 09 ❖/RRD 10 9 8 7 6 5 4 3 2 1

To my own girls,
Emily, Sarah, and Alice,
who never cease to amaze me

It was as if we were being punished for loving the loveliness of summer.

—Willa Cather, *My Ántonia*

I often times think of those days; when it seemed possible that if a man seated on the gable peak of the old sod house by reaching quietly up when a flock of wild geese were flying he could easily reach up and catch them. But now where are our geese; way above the clouds; we can hear them but can't see them.

—Josephine Buchmillar Leber, Dakota pioneer

Contents

The Children's Blizzard

January 12, 1888
The Advance of the Cold Wave

The low-pressure system that drew the arctic air mass down from Canada is shown at 2 p.m. The heavy arcing lines mark the leading edge of cold air at three critical hours. Central Standard Time; pressure readings in inches of mercury.

Prologue

On January 12, 1888, a blizzard broke over the center of the North American continent. Out of nowhere, a soot gray cloud appeared over the northwest horizon. The air grew still for a long, eerie measure, then the sky began to roar and a wall of ice dust blasted the prairie. Every crevice, every gap and orifice instantly filled with shattered crystals, blinding, smothering, suffocating, burying anything exposed to the wind. The cold front raced down the undefended grasslands like a crack unstoppable army. Montana fell before dawn; North Dakota went while farmers were out doing their early morning chores; South Dakota, during morning recess; Nebraska as school clocks rounded toward dismissal. In three minutes the front subtracted 18 degrees* from the air's temperature. Then evening gathered in and temperatures kept dropping steadily, hour after hour, in the northwest gale. Before

---

*All temperatures are Fahrenheit unless otherwise indicated.

midnight, windchills were down to 40 below zero. That's when the killing happened. By morning on Friday the thirteenth, hundreds of people lay dead on the Dakota and Nebraska prairie, many of them children who had fled—or been dismissed from—country schools at the moment when the wind shifted and the sky exploded.

Chance is always a silent partner in disaster. Bad luck, bad timing, the wrong choice at a crucial moment, and the door is inexorably shut and barred. The tragedy of the January 12 blizzard was that the bad timing extended across a region and cut through the shared experiences of an entire population. The storm hit the most thickly settled sections of Nebraska and Dakota Territory at the worst possible moment—late in the morning or early in the afternoon on the first mild day in several weeks, a day when children had raced to school with no coats or gloves and farmers were far from home doing chores they had put off during the long siege of cold. But the deadly quirks of chance went deeper and farther than circumstance or time of day. It was the deep current of history that left the prairie peculiarly vulnerable to the storm.

For nearly all of the nation's short life span, the grasslands at the heart of the country had been ignored, overlooked, skirted, or raced over. On maps the words *Great American Desert* hovered vaguely between the Mississippi River and the Rocky Mountains, and the rest was left blank or faintly labeled Indian Territory. But then, after the Civil War, when the swelling cities of the East Coast settled down to the serious business of industrial capitalism, the Great American Desert was reborn and rechristened. Suddenly this immense expanse of open land was not waste, but paradise—and like paradise, it was free, or all but free. Railroad companies flush with millions of acres of government land grants promised new settlers the sky and sold them the earth at irresistible prices. Under the Homestead Act, the U.S. government gave every comer 160 acres free and clear in exchange for the investment of a small filing fee and five years of farming. The dream of free land let loose a stampede. In the three decades after 1870, some 225 million acres of the continent's heartland were broken, stripped of sod, and planted with crops—more land than had been "improved" in the preceding 263 years of white settlement in the United States. On the last frontier was enacted the greatest human migration the earth had yet endured.

It was late in the day to be an American pioneer. While Thomas Edison was making the first moving pictures in New Jersey, while electric lights shone from Chicago skyscrapers raised on steel skeletons, while Vanderbilts, Carnegies, Morgans, and Rockefellers were adorning their neo-Gothic and Renaissance palaces with the treasures of Europe, homesteaders in Dakota warmed themselves in sod huts at fires of buffalo bones. It wasn't that the sodbusters didn't know that elegant Pullman sleeping cars skimmed over the train tracks at the edges of their wheat fields or that the future price of that wheat depended on tycoons in New York and the number of mouths to feed in Russia. Whether they had come from Europe in the reeking steerage of immigrant ships or boarded converted cattle cars in Chicago, Saint Paul, or New York, they had witnessed with their own eyes the newborn marvels of the industrial world. Someday, they believed, these marvels would be theirs. If they worked hard enough, if their children worked hard enough, the land in time would provide.

And so the settlers of the prairie banked on the future and put their trust in land they loved but didn't really understand. They got down to work so quickly they didn't have time to figure out the vagaries of soil and climate, the cycles of the seasons, the fickle violent moods of the sky. Deprived of both the folk wisdom born of deep familiarity with a single place and the brash abstractions of the new science, the pioneers were vulnerable and exposed. There hadn't been time to put up fences. Children waded into tall grass and vanished. Infants were accidentally dropped in snowdrifts. Infections flourished in the primitive, unsanitary claim shanties.

Coded messages hummed through the telegraph wires strung alongside the train tracks, but settlers' farms were too far from the offices where the messages were received and decoded to do them any good. When the cloud descended from the northwest and filled the air with snow, they had no warning. Unaware of the risk, they wandered out in pursuit of a single precious cow and lost their way between sod hut and barn. Their fuel gave out, their roofs blew off, their animals suffocated. Their children froze to death in the furrows of their fields.

"All around no-one knew of any-one else's predicament," wrote a Dakota pioneer after the storm, "so each acted as he or she thought fit and people survived or died according to their temperament. You can't preach about it. If a young fellow had every penny of his cash tied up in an uninsured herd of cattle . . . what would most of us have done? No-one knew THEN that this was the day which was to be remembered when all the days of 70 years would be forgotten."

One of the many tragedies of that day was the failure of the weather forecasters, a failure compounded of faulty science, primitive technology, human error, narrow-mindedness, and sheer ignorance. America in 1888 had the benefit of an established, well-funded, nationwide weather service attached to the Army and headed by a charismatic general—yet the top priority on any given day was not weather, but political infighting. Forecasters—"indications officers," as they were styled then—insisted their forecasts were correct 83.7 percent of the time for the next twenty-four hours, but they were forbidden to use the word *tornado* in any prediction; they believed that America's major coastal cities were immune to hurricanes; they relied more on geometry and cartography than on physics in tracking storms; they lacked the means and, for the most part, the desire to pursue meteorological research. "[T]he promise of a science of profound interest to the scholar and of vast usefulness to the people is being rapidly realized," wrote explorer and geologist Major John Wesley Powell of meteorology in 1891. "While the science has not yet reached that stage when directions can be successfully given at what hour it is wise to carry an umbrella on a showery day, it has reached that stage when the great storms and waves of intense heat or intense cold can be predicted for all the land in advance of their coming so as to be of great value to all industries of the land. All the discomforts of the weather cannot be avoided, but the great disasters can be anticipated and obviated." Mighty rhetoric—and many believed it. But in truth, when it came to weather prediction, government forecasters in the last decades of the nineteenth century were still relying more on empirical observations and even proverbs of the "red sky at night, sailor's delight" school than on a sound scientific understanding of the atmosphere. Many of the "great storms and waves of intense heat or intense cold" escaped them altogether—or were mentioned in their daily "indications" too late, too vaguely, too timidly to do anyone any good. When it came to "great disasters," they knew far less than they thought knew.

It was the age of confidence. Arrogance was epidemic.

The officer in charge of the experimental indications office that had been established in Saint Paul for the express purpose of predicting blizzards and outbreaks of extreme cold on the prairie did not entirely miss the January 12 storm. He knew before midnight on January 11 that it would snow in Dakota Territory and Nebraska the following afternoon and get colder that night. His indications "verified." But they helped few, if any, people in the region escape or protect themselves. Warnings were not posted in time. No one reading the indications for that day would have guessed that an historic storm was bearing down on them. Those in positions of authority neither recognized nor cared about the forecasting failure. To the extent that knuckles got rapped as a result of the storm, it had to do with sleet-covered sugar plantations in the Deep South, not frozen children on the prairie.

It was the Gilded Age. Disaster meant financial ruin.

Even in a region known for abrupt and radical meteorological change, the blizzard of 1888 was unprecedented in its violence and suddenness. There was no atmospheric herald. No eerie green tinge to the sky or fleecy cirrus forerunner. One moment it was mild, the sun was shining, a damp wind blew fitfully out of the south—the next moment frozen hell had broken loose. The air was so thick with fine-ground wind-lashed ice crystals that people could not breathe. The ice dust webbed their eyelashes and sealed their eyes shut. It sifted into the loose weave of their coats, shirts, dresses, and underwear until their skin was packed in snow. Farmers who had spent a decade walking the same worn paths became disoriented in seconds.

The pioneers of the prairie, even those who had lived there only a few seasons, were accustomed to seeing hail rip open the bases of enormous black clouds and winds of summer fire stream out of the west. They had crouched by their stoves for dark days and nights while winter gales blew without ceasing. They had watched houses get sucked in whirling fragments up the bases of funnel clouds. But nobody had any idea that the atmosphere was capable of a storm like this.

The blizzard of January 12, 1888, known as "the Schoolchildren's Blizzard" because so many of the victims were children caught out on their way home from school, became a marker in the lives of the settlers, the watershed event that separated before and after. The number of deaths—estimated at between 250 and 500 —was small compared to that of the Johnstown Flood that wiped out an entire industrial town in western Pennsylvania the following year or the Galveston hurricane of 1900 that left more than eight thousand dead. But it was traumatic enough that it left an indelible bruise on the consciousness of the region. The pioneers were by and large a taciturn lot, reserved and sober Germans and Scandinavians who rarely put their thoughts or feelings down on paper, and when they did avoided hyperbole at all costs. Yet their accounts of the blizzard of 1888 are shot through with amazement, awe, disbelief. There are thousands of these eyewitness accounts of the storm. Even those who never wrote another word about themselves put down on paper everything they could remember about the great blizzard of 1888. Indeed, it was the storm that has preserved these lives from oblivion. The blizzard literally froze a single day in time. It sent a clean, fine blade through the history of the prairie. It forced people to stop and look at their existences—the earth and sky they had staked their future on, the climate and environment they had brought their children to, the peculiar forces of nature and of nature's God that determined whether they would live or die.

What follows is the story of this storm and some of the individuals whose lives were forever changed by it. Parents who lost children. Children who lost parents. Fathers who died with their coats and their arms wrapped around their sons. Sisters who lay side by side with their faces frozen to the ground. Teachers who locked the schoolhouse doors to keep their students safe inside or led them to shelter—or to death—when the roofs blew off their one-room schoolhouses. Here, too, is the story of the Army officer paid by his government to predict the evolution of the storm and warn people of its approach. In a sense it is a book about multiple and often fatal collisions—collisions between ordinary people going about their daily lives and the immense unfathomable disturbances of weather.

To understand the causes and consequences of these collisions, it's necessary to trace the elements involved to their sources and points of origin. To tease out from the detritus of the past how these particular families happened to find themselves in the path of the northwest wind on that particular day. To isolate the forces in the atmosphere that conspired and converged to create the wind and the deadly cold it carried in its wake. To see those atmospheric forces through the eyes of an Army forecaster who had been trained to fight Indians, follow orders, and apply fixed rules.

❧ ❧

"Everything changes; nothing does," the poet James Merrill wrote in a poem called "After the Fire." The effects of disaster, no matter how extreme, do not last forever. We bury our dead, nurse the wounded, rebuild, and get on with our lives. Today, aside from a few fine marble headstones in country graveyards and the occasional roadside historical marker, not a trace of the blizzard of 1888 remains on the prairie. Yet in the imagination and identity of the region, the storm is as sharply etched as ever: *This is a place where blizzards kill children on their way home from school.* To understand why and how the deadliest Midwestern blizzard happened the way it did is to understand something essential about the history of the American prairie—indeed about the history of America itself.

CHAPTER ONE

Departures and Arrivals

Land, freedom, and hope. In the narrow stony valleys of Norway and the heavily taxed towns of Saxony and Westphalia, in Ukrainian villages bled by the recruiting officers of the czars and Bohemian farms that had been owned and tilled for generations by the same families, land, freedom, and hope meant much the same thing in the last quarter of the nineteenth century: America. Word had spread throughout Europe that there was land—empty land, free land—in the middle of the continent to the west. Land so flat and fertile and unencumbered that a family could plant as soon as they got there and harvest their first season. "Great prairies stretching out as far as one could see," wrote one Norwegian immigrant of the image that lured him and his wife and three sons to America in 1876, "with never a stone to gather up, a tree to cut down, or a stump to grub out—the soil so black and rich that as somebody said, you had only 'to tickle it with a plow, and it would laugh with a beautiful harvest.'" As for the sky above this land, there was no need to worry. Rain, they were promised, would fall abundantly and at just the right times. Winters were bright and bracing, snowfalls light and quick to melt. "Indeed, it may be justly claimed as one of the most beautiful climates in the world," proclaimed a pamphlet written, translated, and distributed by agents of one of the railroad companies that owned millions of the choicest acres of this land, "and one best adapted to the enjoyment of long and vigorous life." And so they came for land, freedom, and hope, some 16.5 million of them between 1850 and 1900, the majority of them never getting beyond the East Coast cities, but many hundreds of thousands, especially the Germans and Scandinavians, ultimately bound for the vast American grassland frontier bordered by the Mississippi to the east and the Missouri River to the west.

Gro Rollag was one of the seven hundred fifty thousand Norwegians who emigrated to America in the nineteenth century. She was twenty-two years old and a bride of several days when she left her family's farm in Tinn in the Telemark region of southern Norway in April 1873. Gro had married a strapping blond boy named Ole, three years her junior, from a neighboring farm. Rollag was his surname as well, since it was the custom in that part of Norway for families to take the names of the farms where they lived. In Tinn there were six Rollag farms scattered through the valley—North Rollag, South Rollag, Center Rollag, and so on—all of them small and niggardly in yields of barley, oats, potatoes, hay. Growing seasons were short this far north, crop failures all too common in chilly overcast summers, fields so pinched that only the most primitive tools could be brought in. "Our honeymoon took us to America," Gro Rollag wrote fifty-six years later with her dry humor, as if they might have chosen Paris or Nice instead. While the truth, of course, was that Gro and Ole left Tinn because the fields of the Rollag farms were being divided into smaller and smaller parcels every generation, because they didn't want to leave their children with less than they had, because in Norway only the firstborn sons inherited the arable valley parcels known as *bonde gaard,* and because Ole was facing five years of compulsory military service.

But it wasn't in Gro's nature to write this in the memoir she titled "Recollections from the Old Days." Nor did she mention how hard it was to leave behind this stunningly beautiful landscape at the beginning of spring—the mountains rising sharply from the shores of a twenty-five-mile-long lake known as the Tinnsjo, the farms clustered on a level shelf of land at the head of the lake, the waterfalls gleaming on the sides of the mountains and feeding streams that merged into the broad Mana River, the red and white farmhouses scattered around the stately white church. Beauty was abundant and free in the countryside of Tinn—but you couldn't eat beauty, and the beautiful farms were yielding less and less while the population steadily grew. But they were comparatively lucky in Tinn. Elsewhere in Telemark the farm fields had become so small from repeated division that farmers had to harvest the hay that grew on the thatch of their roofs and grow vegetables by spreading dirt and manure on top of rocks. It was a sad, haunted country for all its beauty. Men in the prime of their lives built their coffins and stored them inside until they were needed. "It was not a very pleasant thing to look at before you got used to it," recalled one Norwegian immigrant.

Gro Rollag was no beauty, but she was a strong capable young woman with a long face, prominent cheekbones, high forehead, and a kindly intelligent look in her rather narrow eyes. According to family lore, she was not the most conscientious housekeeper because she preferred reading to housework. A love of books and reading ran in the family. Of all the possessions they were forced to sell or leave behind in Norway, what the Rollags remembered with deepest regret was the library they inherited from an eighteenth-century ancestor—lovely old books sold to pay for their passage to America.

Gro and Ole were the first of the family to emigrate, leaving Oslo on April 24, 1873. "We traveled via England and with the Cunard Line from Liverpool," Gro wrote in her recollections half a century later, furnishing precious few details. "We were thirteen days on the Atlantic and landed at Boston. From there we went west in a railroad boxcar. We took a little snack for the journey—a piece of sausage and a few crackers each."

Her brother, Osten Knutson Rollag, was a bit more expansive when he wrote down his own story. Osten explained that their mother, Kari Nilsdatter, had been left a widow in 1862 with three children to support—Soneva, the oldest, was thirteen; Gro, eleven; and Osten himself, eight. It was the custom in that part of Norway for children to work to support themselves right after confirmation—at fourteen or fifteen—so presumably Soneva got a job soon after her father's death, probably as a maid for a neighboring farm family. Soneva seems to have been the family favorite. "She was a more than usually nice person," wrote Osten, "and respected by all who knew her."

Soneva died in 1873 at the age of twenty-four. Her death severed the family's ties to Norway. That same year, Kari sold the farm that her husband had purchased thirty-one years before, Gro married and left for America, and Osten and Kari followed them the next spring. "On the morning of the 15th of May, 1874, we left the home in the valley where my forefathers had lived for how long one does not know," Osten recorded solemnly. "The morning of May 15 began with bright sunshine and the old 'graend' was very beautiful. In the sunshine we saw the new foliage on the birches and the many rushing waterfalls which flowed into the valley. It was very hard to say farewell forever to all of this." He was nineteen years old.

"Among the various arguments for going to America, the strongest was the poverty among the common people where we lived in Norway," wrote a fisherman named Lars Stavig, who left his home in Romsdal on the west coast in 1876. "Also, the hope-lessness of ever amounting to anything and the hard struggle awaiting my boys if they were to remain in Norway."

※ ※

Two families to a wagon—they had agreed on this beforehand. The women would sit on top of the trunks and bags and bedrolls with the smaller children, while the men and older children walked alongside. Of the fifty-three families who loaded the wagons to overflowing that day, Anna and Johann Kaufmann were among the less encum-bered. They only had the two children, a three-year-old named Jo-hann like his father and Peter, a baby who would ride in his mother's arms. Some of their neighbors had five, seven, ten children to look after, mountains of luggage, feeble elderly parents. Until that day, Anna Kaufmann had spent her entire life in the village of Waldheim amid the wide windswept fields of the Ukrainian province of Volhy-nia. She prayed with her family and neighbors every Sunday in the Mennonite church where her father, the Reverend Johann Schrag, served as the elder. The farthest Anna had ever traveled was to neigh-boring villages—Horodischtz, where her husband had grown up, Kotosufka, Sahorez, German-speaking Mennonite settlements whose names have long since fallen off the map.

In a single summer day, all of these villages emptied.

In the weeks before, the fields and farmhouses had been sold to neighboring farmers, the horses and wagons to peasants, the furni-ture and kitchen items to Jews. The women packed baskets with flat bread and sausage and dried fruit for the long journey. The men scraped together enough rubles for the expensive Russian pass-ports. Then came the day of departure. Overnight, Horodischtz, Waldheim, Kotosufka ceased to be the homes of the Kaufmanns, Grabers, Albrechts, Schrags, Preheims, and Gerings. Fifty-three families, some 342 people in all, left together for America late in July 1874.

"Schweizers," these Swiss-German Mennonites called themselves, though their families had not lived in Switzerland for some two hundred years. Because they practiced a different kind of Protestantism from their neighbors, they had been expelled from their farms in Emmental in the Canton of Bern in the 1670s. Rather than baptize their infants a few days after birth, the Schweizers waited until they were old enough to choose baptism as a "confession of faith." They advocated complete separation of church and state and refused to serve in armed forces or fight in wars. For these beliefs, particularly the last, they had been crammed into the prisons of Bern, sold as galley slaves to Venetian merchants, branded, flogged, burned at the stake, and hounded through Europe. From Emmental to the Rhineland of Germany, from Germany to Alsace and Galicia, and then to Poland and Central Ukraine near Zhitomir (west of Kiev), the Schweizers had fled and started over again every few generations—always moving together in groups of families, always settling together in enclaves of villages, always retaining their German language and Swiss customs, always clinging to their Mennonite faith.

They had come to the Polish-Ukrainian border region at the end of the eighteenth century at the invitation of Polish noblemen. It was the same period when thousands of other German-speaking Mennonites, so-called Low Germans, settled farther south in the Crimea at the behest of Catherine the Great. Schweizers and Low Germans alike had been lured to this country by the promise of religious freedom, exemption from military service, the right to own land and to speak German in school and church. And for three or four generations, they had prospered on their small farms between Kiev and Lublin. Hardworking, thrifty, communal, ingenious, the Schweizers had almost uncanny success as farmers. Their flower and vegetable gardens were renowned, their cheese and butter were prized in Kiev and Odessa, silkworms fattened on their mulberry trees, and great swarms of bees pollinated their orchards. But the Schweizers' golden age was short-lived. In 1870, Czar Alexander II withdrew the rights and protections granted by Catherine and inaugurated a policy of Slavicizing the German-speaking Mennonites. If they wanted to remain in the Crimea, they would have to submit to Russian military service and send their children to schools where only Russian was spoken.

Four years later, in the summer of 1874, Anna and Johann Kaufmann and all the families in their congregation piled their remaining possessions into wagons they had borrowed back from the Ukrainian peasants they had sold them to—the first wave of a mass decade-long migration that would bring some eighteen thousand Mennonites from Russia and the Ukraine to America between 1873 and 1883. For Anna, the hardest part was that she would be leaving a child behind, her first Peter, who died at the age of four the year before. It was not callous of Anna and Johann to use the name again when another son was born just months after Peter's death. So many children died in those days that it was customary to keep the name alive with succeeding children. Before the day of departure, Anna's father, Johann Schrag, led a daylong prayer service at the Mennonite church in Waldheim. That would have been Anna's last visit to Peter's grave. A small woman of twenty-four, five years younger than her husband, fair-haired and open-faced, Anna was gentle and tenderhearted and devout. In later years her grandchildren remembered that when she came to greet them Anna always had "a smile on her face and tears in her eyes, which were tears of joy. She laughed and cried at the same time." *"Die freundliche Grossmutter,"* they called her—"the friendly grandmother." Anna wept at Peter's grave, knowing it would be the last time, thanking God that He had given her two more sons, praying that these boys would live longer than their brother.

Before they left Waldheim, the Schweizer families raised their voices in a song of farewell. *"Jetzt ist die Zeit und Stunde da, dass wir zieh'n nach Amerika"* (Now the time and hour are here that we should move to America). Then all bowed their heads and folded their hands in prayer. Supposedly, the peasants from surrounding farms gathered in large numbers and cried as the caravan of swaying wagons rolled by, although one boy remembered an old Ukrainian peasant telling his parents solemnly that their ship was sure to sink, or if it didn't, then they would certainly be killed and eaten by Indians.

⚜ ⚜

It was the logistics of the journeys that the immigrants wrote about in greatest detail. The emotions they either took for granted or were too shy to record, especially the Norwegians, who were famous for their reserve. (There is a Norwegian joke about an old farmer who, in the grip of powerful emotion, once confessed to his best friend, "I love my wife so much I almost told her.") So there is a great deal in the Norwegian memoirs about their heavy trunks and chests, often painted blue, and the difficulty of transporting them from their villages to the train stations or harbors from which they embarked. Osten wrote that he and his mother began their journey to America on board a small steamship called the *Rjukan,* which took them down the long narrow Tinnsjo to the village of Tinnoset at the southern end of the lake, where they spent the first night. From there it was sixty-five miles to Kongsberg, the nearest train station—a long way to haul the chests. After some searching and negotiating, Osten finally found a farmer named Anderson Moen who was willing to take their chests in his wagon. A few yards out of town it became apparent that Moen's horse was so wretched that he could not possibly haul both the chests and the passengers, so the two men walked while Kari rode. They must have been a striking sight in the middle of the road—Osten a muscular young man of medium height, not yet twenty, with reddish blond hair, a mustache just coming in, and blue eyes that "sparkled with intelligence and humor," as a relative wrote; the dignified and artistic Kari, a handsome widow of fifty-two, outspoken, well read, opinionated, a fierce advocate of female suffrage; and the bumbling Anderson, whose nickname was "Bi Litt" (Wait a Little)—his favorite expression. "Anderson was a strong and sturdy fellow," wrote Osten. "I believe that he pushed more than that little horse pulled. It was some fine procession that struggled through the parishes on its way to Kongsberg."

An immigrant named John B. Reese remembered setting out from the mountain town of Opdal in central Norway with a group of families in April 1880. It was a "strange and significant scene," he wrote years later. "Here comes a procession of twenty or more sleds, each drawn by a single small horse. The sleds were heavily loaded with large, blue-tinted chests, as also trunks, satchels and numerous smaller articles of household and family use. Riding on top of these loads are mothers with little children as also a number of grandmothers, the latter upwards of seventy years of age." Reese recalled turning around for a final view of the snowcapped mountains and evergreen forests before a bend in the road swept the familiar landscape from view. Another Norwegian child remembered boarding a ship at 4 A.M. and standing on deck to watch hundreds of people standing at the edge of cliffs waving pale handkerchiefs in the midsummer twilight.

Osten and Kari Rollag had a week in Oslo (then called Christiania after the Danish king Christian IV, who conquered it in 1624) before embarking on the ill-fated steamship *Kong Sverre*. This 310-foot-long two-masted iron-hulled ship was the pride of the Norwegian-owned Norsk Line, with elegant cabins for 75 first- and second-class passengers and accommodations belowdecks for 650 steerage passengers. But the *Kong Sverre,* named for an especially brutal twelfth-century king, lasted only two years—from her maiden voyage out of Bergen on June 29, 1873, until she was wrecked near the entrance to Dunkirk harbor on October 16, 1875. The wreck of the *Kong Sverre* bankrupted the Norsk Line, forcing Norwegian emigrants to embark for America thereafter from foreign ports. For Osten and Kari, the voyage on board this grand vessel was the experience of a lifetime. Osten wrote of the great crowd that gathered at the wharves in Christiania to see the steamship off: "When the ship pulled away from the dock there was waving of handkerchiefs as long as we could see land, and then all stood and sang the national anthem, 'Ja, vi elsker dette landet.'" The ship stopped at Bergen, where she took on more passengers, so that by the time she left for America on June 4, 1874, she carried emigrants from every corner of the country, "from Vestland, from Nordland, from Trondheim, in all 800 people," according to Osten, "all Norwegians."

In the endless days of June there was little sleeping and, for the first- and second-class passengers, much dancing on the decks—at least at first. Then the ship hit the deep swells of the open ocean, and sea sicknesses put an end to the dancing. "The fine ladies who had danced so joyfully during the last days in Bergen lay around on the deck and vomited," wrote Osten with a touch of malice. Of his own quarters belowdecks, Osten mentioned only that he and his mother were shocked to find "nothing more than hard boards—and . . . plenty of lice," but one can imagine the squalor of the unventilated bunk rooms packed with 650 immigrants. On Sunday the faithful gathered for prayers in the morning and again in the afternoon. Somewhere in the mid-Atlantic a powerful storm hit the *Kong Sverre* and "giant waves rolled over the great ship and the water flooded some areas." Part of the rudder broke in the storm and the ship stood idle for two days while the crew repaired it. Fearful passengers begged the captain to turn around and head for England, but he refused. In the event, the weather quieted and they steamed into New York harbor on June 20, 1874, eighteen days after embarking from Christiania.

"We enjoyed ourselves very much on the ship," Osten concluded the saga of his voyage. "The trip was outstanding but there were a few cranks who complained about everything on board and went around with a list in order to send complaints to the company. No one signed it."

❧ ❧

The Tislands, also from the Telemark region, were not as fortunate in their crossing as the Rollags. Of the nine children born to Ole and Karen Tisland, five had died of diphtheria and were buried in Norway. Though their son Andreas survived the disease, he was left deaf and weakened. Andreas was six and a half when Ole and Karen emigrated to America with their three other children. Their crossing was rough. In the course of the voyage, twenty-two children and one adult died. Ole and Karen watched helplessly as Andreas shivered with fever in the unheated steerage quarters. When he died his body was sewn into a canvas shroud with weights attached to either end. The ship's captain read the last rites, and then the bundle was tipped off the side of the ship and into the sea. Some mothers on board immigrant ships kept the deaths of their children secret so they could bury them properly on land. Even burying a child in the strange land of a country they had never seen was better than losing a child's body to the ocean. About one in ten steerage passengers died on board immigrant ships.

The Norwegians journeyed to America on the strength of rumors, railroad company propaganda, hearsay, and letters from friends and relatives, "the America letters," singing the praises of the New World. But the Swiss-German Mennonites, characteristically, wanted to see the country for themselves before making up their minds to emigrate. In the summer of 1873, three years after the Czar revoked their protections, two Schweizer leaders left on a scouting party to America with ten other Ukrainian Mennonites. Shortly after their arrival, the somber black-clad elders managed to secure an audience with President Ulysses S. Grant in Washington, D.C., to request the same privileges and exemptions they had enjoyed under Catherine the Great. Strangely, General George Armstrong Custer attended the meeting and conversed with the Mennonites in German (he had picked it up from his German family). Grant promised nothing—no military exemption, no tax incentive, no guarantee of German schools—but the scouts decided to have a look around the country anyway. They spent the spring and summer touring the republic, traveling by train out to Chicago, Saint Paul, and Duluth, then westward by horse and wagon through the still largely unsettled expanses of Dakota Territory, north into Manitoba, and then south to Nebraska. They liked what they saw, especially the great empty prairies of Dakota. It was July and the unbroken grass looked rich and beautiful and full of promise. No trees to clear, no neighbors to disturb them, abundant sun. In the few homesteads scattered on the river bottoms, they examined the potato patches with approval. By the time the delegation returned to the Ukraine, the Schweizers among them had decided that Dakota was the place.

And so the following July, Anna and Johann Kaufmann and their two young sons began their journey from Waldheim to America in a long caravan of wagons. It took two days to reach the nearest train station at the Ukranian city of Slavuta, some fifty miles east of the border of the Austrian Empire. Most of the Schweizers had never laid eyes on a train before—and there were many prayers offered for their safety. They traveled by train to Brody near the Ukrainian-Austrian border and on to Lemberg (Lvov) in Galicia and then, changing trains, on to Breslau, the principal city of Silesia, where they spent the night on the floor of a spare room next to a beer hall and endured the taunts of drunken patrons. From Breslau they took another train to Berlin and from Berlin to Hamburg, where they found lodging in an "immigration house." A German Mennonite preacher who was invited to pray with the Schweizers at Hamburg left a moving account of the "unforgettable worship service" with some three hundred faithful in attendance. Gathering in the evening in the close quarters of the immigrant house, the congregation began by raising their voices in the migration song, "In all my deeds I let the Lord rule," and then sat in silence as the text of Isaiah 44:24-28 was read aloud: "I am the Lord that maketh all things . . . that saith to Jerusalem, Thou shalt be inhabited; and to the cities of Judah, Ye shall be built, and I will raise up the decayed places thereof: That saith to the deep, Be dry, and I will dry up thy rivers. . . ." At the end, after the preacher "pronounced the blessing of the Lord for the last time in this part of the world," the Schweizer men sang in four-part harmony.

Thus fortified by prayer, the group boarded a steamship at Hamburg and crossed the North Sea to Hull. Then yet another train from Hull to Liverpool. Here they boarded the 4,770-ton 445-foot steamship *City of Chester,* one of the largest ships on the Inman Line, bound for New York. What struck them most about the ship was the fact that all the waiters and cooks were black—they had never encountered people of African descent before. One little boy was convinced that the first black man he saw was "old Nick himself." Traveling in steerage, the Schweizers did not even glimpse the ornate luxury of the first-class staterooms and public rooms above—dining room tables set with linen and crystal, velvet sofas, carved paneling in the saloons. But they were better off than most emigrants. William Inman, the principal owner of the Liverpool-based line that bore his name, was determined that his modern iron-screw steamers provide steerage passengers safe, sanitary passage without "the discomforts and evil hitherto but too common in emigrant ships."

The passage of the Schweizers was not without tragedy. Anna and Johann Kaufmann's baby, Peter, died before the *City of Chester* reached America. For a group as tight-knit and community-minded as the Schweizers, the loss of one child was a loss to all. Anna's father gathered his congregation into a quiet corner of the steerage quarters and led them in prayer for the eternal life of his unbaptized infant grandson. Johann Schrag may well have chosen a text from Revelations, his favorite book of the Bible, to weave into the prayer service. A pious and austere man, even by Mennonite standards, Schrag was quick to see dire signs and portents in the tragedies of life. When the prayers were ended and the last hymn sung, the small body was taken up to the deck and consigned to the Atlantic Ocean.

It was some comfort to Anna to have her entire family on board the ship with her—her four brothers and their wives and children, her two unmarried younger sisters. And there was her three-year-old son, Johann, to look after. After seven years of marriage, after the births of three sons, Johann was all Anna and her husband had left. One small child to bring with them into the unknown reaches of the New World.

❧ ❧

The *City of Chester* arrived in New York Harbor on August 24, 1874, anchoring just off the Battery. A vigorous cold front had pushed through the previous night, dropping temperatures from the mid-80s to the upper 60s and clearing the rank city air. The families gathered on deck to look across the dark water at a city of cobblestone and brick and steep pitched roofs, a hodgepodge of three- and four-story buildings and narrow streets jammed with horses and carts. From the waterfront in those days you could still see the spire of Trinity Church jutting above the rooftops and the granite towers that would soon be strung with the cables of the Brooklyn Bridge. White sails and gray columns of steamer smoke rippled on the water, and the crowded buildings loomed at the water's edge with that ineffable sense of infinite possibility peculiar to New York harbor. Since the *City of Chester* was too big to tie up at one of the wharves that radiated out of the Battery, the ship remained in the deeper water offshore while a smaller boat ferried the passengers and their baggage across.

Family by family, the Kaufmanns and their neighbors from Waldheim and Horodischa—the Albrechts, Grabers, Schrags, Blocks, Gerings, Preheims—marched down the wharf and directly into the low-domed circular building known as Castle Garden.

This curious structure, neither a castle nor a garden, was an early example of creative urban recycling. Built between 1808 and 1811 to fortify the southern tip of Manhattan (hence the name Battery), Castle Garden was reborn as a summer restaurant in 1824; then in the 1840s it was roofed over and converted into an opera house and theater (Jenny Lind, the Swedish Nightingale, sang there for an audience of four thousand); and finally, in 1855, it became the nation's primary immigrant processing center. Over the next thirty-four years, more than eight million immigrants passed through these thick red granite walls on their way to new lives in America.

To the Schweizers, the scene inside Castle Garden looked like pandemonium and sounded like Babel. Immigrants in their heavy woolen clothes filled the rows of benches on the lower level. Overhead hung two tiers of balconies where families who had arrived earlier camped while awaiting clearance. Children shouted and babies squalled. The smell of cheese, rolls, and coffee drifted in from the humid kitchens. Red-faced officials tried vainly to contain and channel the human flood. At the center of the great theatrical rotunda capped by a glass dome stood a dozen representatives of the leading railroads accosting the newcomers with offers of every kind—cheap land out West, easy transportation, temporary lodging while they looked around. A few paces away, an immigration official standing on a kind of rostrum shouted instructions over the hubbub—how and where to get rail or steamer tickets, where to register for employment, how to change money without falling prey to the sharpers and runners and scalpers who were lying in wait outside. Each immigrant was called up for a thorough physical examination, and those with illnesses were removed to a hospital run by the city.

The Schweizers wisely changed their rubles for dollars inside Castle Garden—what few rubles they had left after paying the equivalent of fifty dollars each for Russian passports and the steep fares for the trains and ships they had been traveling on for nearly a month. They had been warned that railway company agents would try to lure them with competitive offers, and they were ready to do business. Before leaving Russia, the families had chosen three leaders to represent the group interests, and these three, with the help of an earlier Mennonite immigrant named David Goerz, who had traveled back to New York to be of assistance, arranged with one of the railroads for a special "immigrant train" to transport them all out to Dakota.

For immigrants traveling alone and without friends or relatives to greet them on arrival, Castle Garden could be a nightmare. John Reese, who was six when his family arrived in New York from Opdal in Norway, remembered that the most terrifying moment of the long journey occurred at Castle Garden. John's parents entrusted him to the care of a servant girl while they went off to arrange for their train tickets. In the milling confusion of hundreds of families speaking strange languages, John wandered off through the "vast spaces" of the Battery and ended up back at the docks. By this time he was sobbing hysterically for his father. At the dock, an immigration official, assuming the child had become separated from his family on board a newly arrived ship, took him out on a ferry to where this ship was anchored. One of the women on board promptly claimed that John was her son. Had the boy not howled in protest, the official would have left him. Somehow John was returned to shore and found his frantic parents.

A harrowing story was told by Finnish immigrants of one of their countrywomen who went into labor just as her immigrant ship anchored off the Battery. The woman was taken to a hospital on shore and forced to leave her baggage and her two-year-old daughter unattended on board the ship. While she was in the hospital, the ship returned to Europe.

"In New York we lost heart again," wrote Norwegian immigrant

Aagot Raaen in her sad and lovely memoir *Grass of the Earth*. "We could not speak the language. We were driven like cattle onto trains that took us to Wisconsin and Iowa. We came from Wisconsin and Iowa to Dakota in covered wagons; we came through a country that had no bridges and no roads; we often traveled for days without seeing anything but prairie. But we again arrived. Empty-handed, we started to work."

The fifty-three Schweizer families stayed in New York for a week during the last week of August, waiting for an even larger party of their fellow Schweizers to arrive from Europe on the *City of Richmond*, another Inman Line steamship. There were some 440 Schweizers in this second group, most of them bound for Kansas, but fourteen of the seventy-three families decided to split off and throw in their lot with the first group. Among these fourteen were Johann and Maria Albrecht, from the village of Kotosufka. There is no record of a prior friendship between the Albrechts and Anna and Johann Kaufmann, though their villages were just a few miles apart in the Ukraine and their families connected by kinship, as all the Schweizer families were. Nor is there a record that the two families were drawn together in New York once the Albrechts decided to join up with the Kaufmanns' group. But there was a strange symmetry in their recent experiences that may well have served as a bond between them. Just as Anna and Johann Kaufmann had lost a son on the ocean voyage to America, so Maria Albrecht had borne a son on board the immigrant ship, a baby boy named Johann after his father, delivered on August 28, 1874, three days before the *City of Richmond* dropped anchor in New York.

For Maria and Johann Albrecht, the birth of their son in the steerage quarters of the immigrant ship came as both a blessing and a terror. Johann at twenty-seven and Maria, twenty-four, had already lost three babies—two daughters and a son—back in the

Ukraine. The Albrechts had been married eight years. They were a small couple, hardly bigger than children themselves, with the round faces, fair complexions, and high, fleshy noses characteristic of their people. Just weeks before, they had sold almost all they owned to Ukrainian peasants and Jews back in Kotosufka; they had spent almost all their money to pay their way across a continent, an ocean, and soon half of another continent. Now, with an infant of three days, a trunk, and the clothes on their backs, they were starting over again on the strength of their faith alone.

The Schweizer group, swelled to sixty-seven families by the addition of the *City of Richmond* party, planned to travel by train from New York, or rather from Jersey City, New Jersey, the metropolitan area rail hub then, to Yankton in Dakota Territory, which was as far west as the train went in 1874. They chose Yankton because the previous year a small Schweizer contingent led by the Unruh and Schrag families had claimed homesteads north of the town. Letters were posted back to the Ukraine describing the open land and the deep, rich soil. The advance party had already built sod houses and plowed small fields. They would help the newcomers get established. Yankton was the place.

The special immigrant train that the Schweizer leaders booked proved to be little more than cattle cars fitted with hard wooden benches—no tables where they could sit down to proper meals, seats barely big enough to accommodate children, no possibility of lying down to sleep, a viciously indifferent crew. When the train stopped to refuel, the crew refused to linger long enough for the immigrants to buy food, and soon they were suffering through what one called "sweltering foodless days when some of us almost perished." A fire broke out in the baggage cars in Buffalo and many of their belongings were destroyed. Chicago, their next stop after Buffalo, was still largely in ruins from the devastating fire of October 1871. At Sioux City, Iowa, the immigrants rebelled. The men descended en masse and marched down the street searching for food. The engineer whistled repeatedly and finally in disgust started the train rolling, but the men paid no attention. When they returned to the station with their provisions, they found the train waiting for them: The conductor, his bluff called, had backed up. Mennonite stubbornness and communal action had prevailed.

It was afternoon by the time the train shuddered to a stop at Yankton on the Dakota side of the wide Missouri. No accommodations could be arranged for so large a party. The families spent their first night in Dakota sleeping out under the stars using their dusty bags for bedding. The first task at hand the next day was to find the Unruh/Schrag settlement north of town in the bottom lands of Turkey Creek. A delegation set out from town on foot. The trodden earth, milled lumber, and sawdust of Yankton made but a small brown scratch on the prairie. After a few dozen paces the sea of summer grass, as deep as their waists and as wide as the horizon, closed around them. They followed an old Indian trail that had been deepened and rutted by the wagon wheels of pioneers. Prairie chickens scurried out of the grass ahead of them and quail flew up in little panicked explosions. When the flapping of wings died away, the silence was absolute but for the drone of mosquitoes and the soughing of wind through the brittle blades. The men sweated in their woolen traveling clothes and by noon they were sunburned—the sun literally scoured the fair skin off their faces. They yearned for a bit of shade. Miles away they could see the pale green of cottonwoods shadowing the banks of some creek or stream, but otherwise there was not a tree or even so much as a decent shrub to break the flow of grass. The few dugouts and sod huts they passed only made the prairie look vaster and lonelier. The first night out, one man swore they must be approaching a thickly settled part for the horizon twinkled with lights: It turned out he had been gazing at fireflies playing in the breeze over the next rise. The one reassuring feature of this strange land was the presence of many small lakes and potholes dotted with waterfowl and thick with tall reeds. It was customary for Schweizers to settle by ponds or streams, so they took the bits of blue water in the midst of all the tawny grass as a good sign. And as farmers they understood instinctively the immense agricultural potential of the prairie. "Where there is such an abundance of grass," one of the men remarked, "grain can also grow there, if the ground is worked up." By the time they returned to their families in Yankton, they were ready to stake their claims.

The Indian trail through the grass seemed less alien the second time they walked it. About thirty-five miles north of Yankton, the men fanned out across the prairie. Wherever one of them sighted a small lake or stream, he would drop his hat or coat as a sign that this land was taken. This was how the Kaufmanns, Albrechts, Grabers, and their fellow Schweizers, eventually some sixty families in all, came to settle around Freeman and Marion and farther out around Turkey Creek and Swan Lake. The families that still had enough money left to buy a pair of oxen and a wagon and a few boards of lumber moved out to their claims at once, while those who were too poor stayed behind in Yankton and picked up any paying jobs they could get to earn the cash they needed to homestead. Fifty cents a day was the going wage. The ride out over the prairie behind the expensive but often barely broken oxen was what the women recalled with greatest horror. "Oh such a ride!" one woman wrote years later. "I was so afraid the oxen would run away (no lines). There we sat, flat on the lumber load—nothing to hang onto—the wagon and lumber swaying this way and that way and we poor things slipping and sliding every old way trying to hang onto our kids and our belongings. . . . No bridges on the creeks and so much water in those days." Night fell before they reached the claims, so they made camp near a small lake, using the pieces of lumber for shelter. Some of the younger men who had brought along shotguns from the Ukraine brought down a few ducks as they flew up from the lake—their first taste of wild American game.

❧ ❧

Gro and Ole Rollag, the newlyweds, had chosen Decorah in Winneshiek County in the northeast corner of Iowa on the basis of a sheaf of letters from Norwegian pioneers who had come before them. No mention of the long harsh winters, no word about the grasshoppers that devoured the crops or the prairie fires that consumed the grass in waves of flame. It was all free land, virgin soil, fertile loam, bumper harvests of wheat. So Gro and Ole boarded a train in Boston as soon as they arrived in America and traveled west to Iowa in a boxcar. When they got to Decorah they put up at the house of a relative from Norway named Abraham Jacobson. It was Jacobson who let them know that the "America letters" were not entirely accurate or up to date. The Iowa soil was indeed superb, but the free land was gone, all claimed and proved up by earlier arrivals. Gro and Ole stayed on with the Jacobsons, picking up what work they could in Decorah while they looked around for something better. From Abraham's son Nils they heard about a nice unclaimed stretch of prairie with many creeks and small rivers clear across the other side of the state, just north of the Iowa state line where Minnesota and South Dakota come together. Sioux Falls, situated a few miles west of this country at a promising spot on the Big Sioux River, looked like it would become the principal "city": It already boasted a population of 593, and new buildings were going up at a clip. Come spring, Nils was going to move out there and other young Norwegians from Decorah planned to go with him. In May of 1874, almost exactly a year after they arrived in America, Gro and Ole threw in their lot with this small band of Norwegians and headed west to Rock County, Minnesota.

After borrowing seventy-five dollars from Ole's sister, the Rollags had enough money, just, for the essential outfit of the pioneer—a yoke of oxen, a wagon, and a couple of cows. "Everyone warned us against going west because the grasshoppers were so bad," Gro wrote years later. "But we went anyway." It took the party three weeks to cover the 260 miles of prairie land between Decorah and Rock County. They had almost arrived when they saw the first signs of grasshopper devastation around Jackson on the Minnesota side of the Des Moines River. "They sat thick in the trees by the river," wrote Gro. "Many farmplaces along the way were abandoned, and we came to one hut where the people said, 'Next year you'll go back.' That was hardly encouraging, but we went on toward the goal of our journey, Rock County, and arrived there in June."

Most of the land in Rock County was taken as well, but the Rollags managed to secure a homestead in section 13. "So we had to begin to become farmers," wrote Gro. "It didn't look very good. Only prairie all around and nothing to begin with." They dug a cellar and laid some boards over one corner to shelter their bed. For cooking they had a small stove that they had purchased in the nearby town of Worthington—"so small that we could bake only three loaves of bread in it," recalled Gro, "and when we had to use hay for fuel it looked dismal enough." Ole spent the first summer breaking sod. In the fall Nils helped them build a sod house; they roofed it with some four-by-fours on which they piled more sod to serve as shingles. That first year they lived on a dirt floor.

Osten and Kari came out to join them the following March. Like Gro and Ole, Osten and Kari had spent their first years in America with relatives and then moved farther west when they found out how expensive the land was in the settled parts of the Midwest. Osten wrote in his memoirs that he and his mother journeyed out to Rock County by train, arriving at Worthington on March 2, 1875: "Worthington was then a very lively place, with new houses being built everywhere and the sound of hammering was heard from many places. . . . You could see a forest of ox-carts everywhere." Ole came to meet the newcomers in town with his team of oxen, Spot and Dick, and an empty wagon that they could fill with their purchases—lumber, stove, beds, kitchen utensils. It was at the lumberyard that the brothers-in-law ran into a spot of trouble. Ole brought the oxen to a halt on a patch of gravel that had been scraped clean of snow, and the men loaded the wagon. But once it was loaded, "the oxen could not get off the spot and remained stuck," wrote Osten. "And then the train came. It came from the east. You can believe that the oxen became lively because they certainly had not seen a train before. They made a big hop and got out of the gravel and went at a run through the streets and we ran after them as best we could. It was fun for the townsfolk who saw and enjoyed the foot-race. We had to go back to the hotel to get Mother but the oxen were not willing to turn around so we had a hard time in getting them to do so."

Things went a bit more smoothly once they got the oxen out of town. Gro and Ole's homestead was almost fifty miles from Worthington, and it was a long trudge in early March with the prairie still deep in snow. Finally, well after dark on the night of March 9, they arrived at the couple's sod house—"14x16 feet with a dirt floor and straw roof," Osten remembered, "but it was good and warm." The next day a terrific snowstorm blew in and raged for four days over the open prairie.

If Osten and Kari were discouraged by their first taste of pioneering, Osten made no mention of it in his memoirs. Nor did Gro in her recollections permit herself more than a mild grumble: "We were young," she wrote, "and didn't lose courage because everything looked so gloomy." The country was "pretty wild and desolate when we arrived," wrote Osten. "No trees were to be seen for many miles. Everywhere we saw small sodhouses which were the first dwellings people had. Large groups of Indians passed through, but they did us no harm. We were 49 English miles to the nearest railroad, and it took four days with oxen to go there and return."

The Rollags were born to be pioneers. Ole was strong and strapping, big and loud and booming with optimism. Gro, despite her penchant for reading, was a good manager who kept a level head in times of danger. Still, they quailed before the extremes of weather on the prairie, especially the blizzards that swept down from the northwest with unbelievable speed. Norway had cold winters and plenty of snow, but nothing like the American heartland. The Rollags and the Norwegian families who settled near them were not prepared for the severity of prairie winters. No one was.

※ ※

Johann Friedrich Schlesselmer left his village near Hannover, Germany, in the late 1870s to start over in America, settled in Nebraska, and promptly lost his pittance prospecting for coal south of Seward. He was still a young man when he died of smallpox in 1882, a year after he was naturalized, leaving his wife, Wilhelmine, to manage alone with a house full of children. The Schlesselmers' five-year-old daughter Lena contracted smallpox at the same time as her father. She survived, though the disease left her face permanently scarred. Hard luck plagued the family, especially the little girl. Over the next five years Lena's mother, Wilhelmine, remarried twice, bearing children with each husband (eventually there would be eleven children, eight of whom survived). It was shortly after the third marriage, to a fellow German named Wilhelm Dorgeloh, that Wilhelmine decided—or agreed—to farm Lena out to a new home. The Dorgeloh household out on the prairie near Milford, south of Seward, was crowded with Wilhelmine's latest crop of babies, and naturally Wilhelm Dorgeloh preferred his own children to those from the previous marriages. Lena was large for her age, strong and quiet. With those smallpox scars she would be unlikely to marry early, if at all. Lena must go.

And so, in August 1887, Wilhelmine took her eleven-year-old daughter to live in a hilly section south of Seward with Wilhelm and Catherina Woebbecke, relatives from her native village of

Herkensen in Germany. It was an odd stretch of country—the one place for miles around where the grassland was crumpled into slopes and deep ravines choked with wild plum and willow. Townspeople in Seward (a growing prairie town that boasted three rivers, two rail lines, and a flour mill) had taken to calling this district the Bohemian Alps because of all the Czech and German immigrants who were moving in and buying up the irregular terrain that no one else wanted. The Woebbecke house was small—only one room for the family and a second room used as a woodshed—and they had three little children of their own underfoot. But Wilhelm and Catherina agreed to take Lena in because they needed help with the farm chores and Lena was built like a worker. From then on she was known as Lena Woebbecke. Lena milked and herded the cows and carried water and minded the children when Catherina was busy. After the harvest was done, she started school.

The district 71 schoolhouse was only half a mile away, the minimum distance between just about everything on the prairie, but the brush in the ravines and the steep hill beyond hid it from the Woebbecke place. In fact, the land was so deeply creased here that you had to be practically upon the hip-roofed frame school building before you could see it. Since coming to live with the Woebbeckes in August, Lena had rarely laid eyes on the school. The German Lutheran church they attended was in the other direction, and aside from church on Sunday morning, Lena never went anywhere except where the cows strayed. The district 71 schoolhouse might have been in another country.

Before the fall term started, Wilhelm and Catherina warned Lena that no German could be spoken in school. Only English. Sometimes children were beaten if they spoke a foreign language. The teacher might even change her name. Woebbecke might be too hard for Stella Badger to say.

Lena walked to school alone each morning and back to the Woebbecke house alone each afternoon. Wilhelm had shown her the quickest route—behind the barn, up the hill, down into the ravine, and then up again to the crest where the schoolhouse sat at the edge of a plowed field. Rain or shine, she carried a pail with her lunch. She was always careful to bring the pail back to the Woebbecke place every afternoon. She knew there would be words if she forgot it.

Lena was shy and silent around the strangers. No one paid much attention to her. No one worried about her cutting through the ravines by herself. All day long in school she seldom spoke. As fall dwindled down to winter, she had still made little progress in English. When the bitter cold weather set in, Lena went to school less and less often. Even a light snowfall filled the ravines with drifts and made them impassable. It was rough country for a child to be walking through alone.

It was unfortunate that the Schweizer families arrived on the prairie too late in the year to plant a crop. It was the end of September, nearly autumn, and the nights were already frosty. In any case, there was no time for plowing or planting. It was all they could do to tear enough sod off the prairie to make shelters for themselves. It took half an acre of Dakota sod for a decent-sized sod house. The soddies leaked when it rained ("I would wake up with dirty water running through my hair," wrote one pioneer), gophers and snakes sometimes popped from the walls, dirt got ground into clothes, skin, and food, but they kept the families alive and relatively warm when winter arrived—which happened far earlier and far more savagely than any of the Schweizers had anticipated.

Because of the way the homestead laws worked, the families had to spread out in 160-acre squares checkerboarded over the land rather than live next door to each other in villages with long narrow fields radiating out around their houses, as they had done in the Ukraine. But whenever possible, friends and relatives took adjoining claims. This is how Anna and Johann Kaufmann and Anna's father, Johann Schrag, and her two unmarried sisters came to live on adjoining sections in a township of Turner County with the hopeful name of Rosefield. Freni Graber, who had grown up in Waldheim where Johann Schrag preached in the Schweizer church, and her husband, Peter, took a claim nearby, as did Johann and Maria Albrecht—near enough to see each other's houses, once they were finally able to afford the wood to build houses; near enough to send their children to the same school, once they had the time and money to build a one-room schoolhouse more or less equidistant from their homes at the western edge of section 26.

Peter and Freni Graber were old to be starting over in Dakota with a house full of children—Peter was forty-four and Freni thirty-six when they emigrated with their seven children. Their oldest, seventeen-year-old Joseph, was big enough to plow and plant and would soon marry and start his own family; Anna and Freni, the two oldest daughters at twelve and ten, respectively, looked after three-year-old Peter and baby Johann, not yet two, when their mother needed help. But still it was rough for the Grabers to make a life in this new land. That first winter was long and bitter, with many days below zero and only twisted hay for fuel. For days at a time, the Grabers—the nine of them packed into their sod hut— could not get outside because snow had drifted over their door. When the cold overwhelmed the feeble heat that came from their stove, they stayed in bed all day, children and parents pressed against each other for warmth. Food was scarce, often just corn bread and melted snow to drink.

Freni Graber must have stinted herself that winter in order to keep her children alive. She died in the spring, on May 25, 1875. Peter was married again three weeks later, to Susanna Gering. Susanna was twenty-four years old, only seven years older than her oldest stepson Joseph. Little Peter, now four, would soon have a slew of new half brothers and sisters, eight in all, born every other year or so. The Grabers prized all their children as a gift from God, even when they barely had enough to feed them.

The first Dakota winter took its toll on Anna and Johann Kaufmann as well. All winter long as the supply of food dwindled away, Anna looked at her son, Johann, her only child, grow thinner and more transparent. There were many days when they got by on burnt flour soup—flour scorched in a pan and then mixed with water, salt, and pepper. A poor diet for a growing child. A poor diet for a baby, if they had a baby to feed. Even had Peter survived the crossing to New York, he surely would have perished that first terrible winter. Would it have been worse to chisel a grave for the child in the frozen ground under the sod than see his body tossed into the sea?

Anna had been raised by her father to thank God for the blessings of life and never to complain. Somehow she and her husband and their son endured the Dakota winter and the boy Johann lived to celebrate his fourth birthday in 1875. That same year, their first full year in America, Anna gave birth to another son. They chose the name Jacob from the Old Testament. But the infant died before the year was out, as the two Peters had died before him. Four births now, and still Anna and her husband, Johann, had but the one child to protect and somehow feed through yet another winter.

Johann and Maria Albrecht felt blessed that their baby son, Johann, the child born on board the *City of Richmond,* survived that first cruel winter—and blessed again that all three of them somehow lived through the even crueler first spring. By the time the snow was melting off the prairie grass in April 1875, the Albrechts and their fellow Schweizers were close to starvation. They were only saved by a donation from the Mennonite Committee of Relief, funded by earlier Mennonite settlements back east, that enabled the Schweizers to buy two thousand sacks of flour. But at least they had the land. With land they could plant and harvest and finally have food of their own. But when it came time to plow in April, Johann discovered that his oxen were too weak to break the thick prairie sod—they, too, had nearly starved to death that winter. So Johann and a neighbor had to team up and hitch their four oxen to a single plow, taking turns breaking each other's fields. By working every daylight hour they managed to plow up half an acre a day, perhaps three-quarters on a good day. Geese stood unperturbed a few yards away and watched the men work. Johann seeded by hand, scattering fistfuls of wheat every two to three steps as he walked his field. He planted rye in the next field, and buried cut-up potatoes in the black earth. Like his neighbors, he put his faith in God and the bounty of the land.

The Homestead Act, signed into law by Abraham Lincoln in 1862, was the first color-blind, sex-blind equal opportunity piece of legislation on the American books. White or black, male or female, foreign born or native born, it made no difference. As long as you were twenty-one or older, could muster $18 for the filing fee, and lived on the land and farmed it for five years, 160 acres was yours. The one group the Homestead Act privileged was the military. Those who served in the Civil War had a year stricken from the five-year residency requirement for every year of service in the Union Army.

For Benjamin Shattuck that meant that just two years after he and his family moved out from Ohio, the 160 acres of prime Nebraska prairie belonged to him. Born in the rich rolling farmland of eastern Ohio in 1835, Ben was twenty-six years old and single when he enlisted in the seventy-third Ohio Volunteer Infantry on November 16, 1861, seven months after the war began. He was assigned to Company B under the command of Second Lieutenant Thomas W. Higgins, and he drilled through the cold, wet winter months along with hundreds of other raw recruits at Camp Logan near Chillicothe. By the end of January 1862, the

Seventy-third Ohio was considered battle ready and the men boarded trains bound for West Virginia. Their first taste of action was a forced march of eighty miles over mountain roads in a winter storm. Near Moorfield, on the South Branch of the Potomac, they were ambushed at night by Confederate snipers as they stood warming themselves at roadside campfires. The next day, the Seventy-third came under Rebel fire again while trying to ford the storm-swollen Potomac and take Moorfield. Eventually the Union soldiers prevailed and briefly held the town before retreating back up the river.

Disease ravaged the green regiment in the aftermath of this first battle. Many died in the mud and snow. Whether Ben Shattuck was among those who fell ill during those first bitter weeks of campaigning, we do not know. But he did survive. On March 20, 1862, he was promoted to the rank of corporal. It was sometime during this first year of his service in the Union Army that Ben "converted" to Christianity, as an awakening of religious fervor was termed, and joined the Methodist church—the Methodist Episcopal Church, as it was known then.

Ben served with the Seventy-third Ohio in some of the bloodiest battles of the war, including the disastrous Second Battle of Bull Run at the end of August 1862, in which 147 of the regiment's 310 men were killed or wounded and 20 taken prisoner, and the humiliating Union defeat at Chancellorsville the following spring. Though Chancellorsville ended in confusion and retreat for the massive Union contingent under General Joseph Hooker, the Confederate Army paid dearly for its victory. Robert E. Lee sustained some thirteen thousand casualties during the campaign (about 22 percent of his army) and lost the charismatic General Stonewall Jackson, mortally wounded by accident by his own men while returning to the Confederate lines at night. By chance, the Seventy-third Ohio was positioned away from the worst of the fighting and they emerged from the engagement relatively unscathed. In all,

Union casualties came to more than seventeen thousand men during those few days in April 1863.

At noon on July 1, 1863, the Ohio Seventy-third arrived at Cemetery Hill overlooking Gettysburg, Pennsylvania, and for the next three days they endured the almost ceaseless fire of Lee's army. During the few hours at night that the guns and cannons were silent, the Ohio men lay shivering on the ground, listening to the cries of the wounded and dying on the field. "It was the most distressful wail we ever listened to," wrote Samuel H. Hurst, the regiment's commander.

The climax of the battle came on July 3. Early that day the Ohio men were driven back at the Emmetsburg Road, but eventually they advanced as the Union forces succeeded in breaching Lee's line.

Sometime in the course of that day Ben Shattuck, now a sergeant, sustained a bullet wound in his right leg and was taken prisoner by the Confederate forces. For the next eighty-three days he was held at the Confederate prison camp on Belle Isle, a low-lying island surrounded by rapids of the James River near Richmond, Virginia. There were no permanent barracks for the prisoners, only tents, and food was so scarce that prisoners were reduced to gnawing on maggoty bones and stealing the boots off dying fellow soldiers and selling them for food. "All other thoughts and feelings had become concentrated in that of hunger," wrote a Union prisoner. "Men became, under such surroundings, indifferent to almost everything, except their own miseries, and found an excuse in their sufferings for any violations of ordinary usages of humanity." Every day, fifteen to twenty-five prisoners died. Their corpses were wrapped in canvas and tossed into holes in the ground just outside the prison. Many on Belle Isle were forced to sleep on the ground without shelter and died of exposure; many froze to death in the tents.

"Can those be men?" the poet Walt Whitman wondered when

he saw a group of Union soldiers returning from Belle Isle. "Those little livid brown ash streaked, monkey-looking dwarves?—are they not really mummified, dwindled corpses?"

After nearly three months, Ben was released from Belle Isle, possibly in an exchange for Confederate prisoners. The wound in his leg would bother him for the rest of his life. During his final fifteen months of military service, Ben fought with General Sherman's forces in the siege of Atlanta. He watched the city burn in November of 1864 and he marched with Sherman to the sea. On New Year's Eve of 1864, Sergeant Shattuck's term of service expired and he was mustered out of his regiment.

Details of Ben's life become sketchy once he left the Army. On March 1, 1866, he married Sarah Jane Targe—she was twenty-two years old, he thirty-one, old in those days to be starting a new life. Two years later, on April 8, 1868, the Shattucks' first child was born—a girl they named Allie Etta and called Etta. Eventually the couple had six other children, two of whom died in childhood. It's unclear when or why the Shattucks moved west to Nebraska and how many places they lived in before they settled near Seward, the county seat just west of Lincoln named for President Lincoln's secretary of state William Seward. There is a record of an Ohio-born Shattuck living in Iowa in the early 1870s, though this may have been a cousin. We know for certain Ben and Sarah Shattuck were living near Seward in November 1882, because that's when they joined the town's Methodist Episcopal Church. Whether the war injury made it difficult for Ben to work successfully in the fields, whether he was unlucky, unsteady, haunted by memories of Belle Isle, or just a poor farmer is uncertain—but it's clear that Ben struggled and often failed to support his large family by farming, even though the soil around Seward is good and the climate favorable. Like many another hapless sod farmer, Ben Shattuck decided to pull up stakes and try his luck elsewhere—farther out on the prairie. On March 15, 1885, Ben and Sarah withdrew from the

Seward Methodist church, and around that time the family moved north to Holt County, a bleak, flat, arid region on the edge of the Nebraska sand hills. "B. Shattuck" is listed in an 1886 census as a farmer in the town of Atkinson, Holt County, but once again he failed to raise a crop sufficient to feed his family.

We know these details because, when disaster struck two years later, the Shattucks and their oldest daughter, Etta, briefly became celebrities. The newspapers that told of Etta Shattuck's plight in excruciating detail for a few frigid weeks in January and February 1888 all mentioned that the girl, not quite twenty years old, was the sole support of her parents and four siblings. A Methodist "convert" like her father, a plain young woman with a full square face and brown hair that she parted in the middle, Etta kept the family going on the twenty-five dollars a month she earned teaching school at the one-room schoolhouse in the Bright Hope School District in Holt County. The newspapers loved it: a wounded Civil War veteran, a devout teenage schoolteacher, a terrible act of God. The story ran on front pages for weeks.

Between Ben's service in the Seventy-third Ohio Volunteer Infantry and the blizzard that took Etta, the record of the Shattuck family is dim. Only war and natural disaster have secured them a few lines in history.

❧ ❧

S. F. Huntley and his wife, Abi Townsend Huntley, ordained ministers both, came west to bring the word of God to the Dakota prairie. Abi was forty-one years old and eight and a half months pregnant with her fourth child when the family arrived in the town of Plankinton by train on April 5, 1883. A small straight-backed woman with a long, narrow mouth and sparse fair hair pulled back and tucked neatly behind her prominent ears, Abi brought with her a degree from the Whitestown Seminary in Oneida, New York, a steady devotion to the Quaker faith she had been raised in, crates of precious books both religious and secular, and very little else. At the Plankinton train station the Huntleys hired a young boy and a pair of old, tired horses to drag them, their children, and their few earthly possessions the score of miles to the new-fledged town of Wessington Springs. The horses were so broken down that a team of oxen passed them before evening drew in.

The Reverend S. F. Huntley, though possessed of a sturdy upright frame, a prominent forehead, and a neat goatee, was a bit thin in the shoulders and broad around the middle for a pioneer. By the looks of him, he had more grooming than muscle. Certainly he had more education than most of the men swarming into Dakota in those years. S. F. Huntley prided himself on a degree from Cornell University and his service in the Civil War with Company B of the 152nd New York Volunteer Infantry. For a man of his quality and experience, this move to Dakota was to be viewed as something akin to missionary work. He and his wife intended to see church life established and rooted on the godless frontier. He would found a Congregational Church, she would start a Quaker meeting. Though members of different denominations, they traveled together with that joint mission. The earthly possessions they bore with them were few, but that didn't matter. The Reverends S. F. and Abi Huntley carried their fortune in their heads—and their hearts. For the rest, God would provide.

They arrived in Wessington Springs at night with no claim or kin—just the name of a fellow preacher who had advertised for a new colony. The preacher had no room to put them up, but he showed them the way to the house of a kind stranger. The Huntleys crossed a gully and ascended a series of low hills and finally, with night deep around them, they saw a light burning in the window. It was here, six days later, that Abi, far past her first youth and in an alien primitive country with no home to call her own, gave birth to another daughter. Three daughters and a son now, and all that they owned piled on a wagon.

By June, S.F. claimed a quarter section by squatter's rights, surveyed the land, and filed the papers. That summer the preacher managed to bust two acres of sod and put in a vegetable garden. He built his family a house, really not much more than a shack ten feet by twelve—too small for the six of them, so he cut sod for the walls and roof of an additional room and a shed for storing fuel. A couple of years later, the Huntleys homesteaded another quarter section nearby and built a proper frame house, but still they kept the soddie. When the coldest weather hit each winter, they moved back to the old sod house. Truth be told, for all the dirt and mess of walls made of prairie turf, it was easier to keep a soddie warm.

Settlers came fast, many of them Quakers like Abi. When it was time to name their stretch of Jerauld County prairie, the Quakers prevailed and the name they settled on was Harmony Township. It was a blessing for Abi that a Quaker meeting was organized so soon and she did whatever she could to help. S.F. preached at the Congregational Church, and it pleased him that so many of their neighbors attended regularly every Sunday. A one-room schoolhouse was built not far from their home, less than a mile to the west, and the three older children—Mary, Ernest, and Mabelle—walked out there together when the weather permitted. Miss May Hunt was their teacher, an estimable young woman and, S.F. was pleased to note, a member in good standing of his church.

It made Abi uneasy, this turning the children loose to walk the prairie, what with snakes in the tall grass in the warm weather, and standing water in the gullies, and fierce winds whipping up when you turned your back—not like the weather she was used to in the hilly woods of central New York State. But as her husband often said with pride, Abi Townsend Huntley was not one to fret or complain. Never had she uttered a word of discontent since they moved out. And anyway, it was a comfort to them both that there were so many other families in Harmony Township these days. The Reverend Huntley stood at their front door one morning and, gazing out to the north and east, counted eighty-three houses where four years ago there had been nothing but rolling grass and a couple of sod huts.

Plenty of folks to look out for Mary, Ernest, and Mabelle Huntley as they walked back and forth to Miss Hunt's school.

⚘ ⚘

William Clark Allen went west with his family in 1881—not to bust sod on a homestead or to build a church where they could worship as they pleased or to escape hopeless poverty, but to try something new and large and clean in a town that barely existed until they got off the train and built it. The bearer of a proud Yankee name (family legend claims Ethan Allen as their ancestor— mistakenly, as it turns out), a man of enterprise and gumption with piercing blue eyes, wavy salt-and-pepper hair, a high round forehead, and the flowing walrus mustache of his era, William Clark Allen—W.C., as he styled himself—managed to be a man of both adventure and substance. He was born in 1845 in the Great Lakes town of Newcastle, across from Rochester on the Canadian side of Lake Ontario, moved west to Wisconsin as a young man, went to work as a lawyer (no need for law school in those freewheeling days), married, fathered two sons, lost his wife, married a schoolteacher named Edna Jewett whose family had emigrated to the Midwest from upstate New York, and moved again, west again, this time to Minneapolis. In November 1879, when W.C. was thirty-four and Edna twenty-eight, they had a child whom they named Walter, so now there were three boys—Hugh, who had a clubfoot, William, who turned ten that year, and Edna's new baby.

By all appearances the Allens were doing fine in the burgeoning, relatively civilized river town of Minneapolis. But they were restless. Though they had a house well stocked with children, books, furniture, and piano, W.C. and Edna decided that Minneapolis was not their fate. They must keep moving west, always west. They would resettle on the virgin prairie of Dakota Territory. For the "opportunities," they said.

And so in the summer of 1880, W.C. made a scouting trip. He got as far west as the winding, shallow James River—then called the Dakota River—that spools south through the prairie before joining the wide Missouri at Yankton and decided he had come far enough. W.C. claimed his homestead at an outpost known as Yorkville (prominent enough to house Brown County's first post office before vanishing from the map), and then he had a look around a townsite a few miles east that officials of the Milwaukee Railroad had designated Groton. Nearby towns, or rather train stops that were expected to become towns, were called Aberdeen, Bath, Andover, Bristol, and Webster—every ten miles or so another reminder of the glorious Anglo-Saxon heritage. At Groton, W.C. bought a building lot for one hundred fifty dollars on what was destined to become Main Street, the first lot sold in town, and got to work. W.C. must have been a man of considerable energy and pluck because by the time he returned to Edna and the children in Minneapolis at the end of the summer of 1880, he had managed to construct a simple but ample three-room frame house of milled lumber, shingled roof, and glass windows—four tall windows running down the sides of the house and one by the front door looking out over the mud of Main Street. This was the first frame house in Groton.

In July 1881, W.C. returned with his family and furniture to take possession. It was a memorable trip. The floods that resulted from the sudden late spring melting of the thirty-foot drifts that had accumulated during the long winter had turned the James River into a giant lake stretching across the breadth of what is now South Dakota, and it was touch and go whether the tracks would be clear in time for the Allen family exodus. In any event, the train they boarded in Minneapolis on July 14 was the maiden voyage of the H & D Division of the Chicago, Milwaukee, Saint Paul and Pacific Railroad—"the Milwaukee," as it was commonly called—bound for Groton, which was then (briefly) the end of the line. The Allens, like the Kaufmanns and their fellow Schweizers, traveled in an immigrant car—and they were just as dismayed at how crude and uncomfortable it was. Edna Allen recalled that a cow and team of horses were stabled at the front of the car, the family's belongings (including their piano and one thousand carefully chosen books) were stacked at the rear, while she and her husband and the children occupied rough wooden seats at the center. It was "terrifically hot," so they traveled with the side doors slid back for air. Walter, not yet two, was ill and fretful and had to be held in his mother's arms the entire trip. Hugh and Will, though old enough to look after themselves, made their stepmother's heart constrict whenever they got too close to an open door. The sea of waving grass outside was unbelievably green and tall and lush from the melting off of the deepest snows ever measured on the prairie.

For some reason the train went only as far as Andover, a stop shy of Groton, and the Allens had to complete the final leg of their journey by horse cart. Arrival afforded them no rest. The town, though surrounded by miles of drowsing prairie, was feverish with activity. Every lot lining Main Street was a building site—there was so much construction going on that just the scrap wood lying around the new houses and stores supplied the Allens with all the fuel they needed for their cookstove. Practically overnight the prairie boomtown had a feed and coal store, a blacksmith shop, a building contractor, hotel, lumberyard, and a business mysteriously identified as a "purveyor of liquid hardware." By September 1881, just weeks after its founding, Groton boasted two rival newspapers: the *Mirror* and the *Groton News,* which prided itself on beating the *Mirror* into print by two days.

W.C. was right there in the thick of it all. Upon arrival he began practicing law out of his home, but he soon diversified into trade, local politics, and civil service. He teamed up with one Frank

Stevens to open a hardware store and a lumber, harness, and tin emporium near the post office, which was also under his control, first as deputy and then head postmaster (Edna, meanwhile, ran the original Yorkville post office during the summer months). W.C. later went into the real estate business, became the town's "police justice," and, according to one of the local papers, "fixed up a court room at the rear of the post office," where he handed out stiff fines to some of Groton's leading citizens when they broke the town's prohibition ordinance.

He sounds a bit like the Wizard of Oz—a comparison that springs readily to mind since thirty-two-year-old L. Frank Baum took up residence in nearby Aberdeen (two stops west on the train) a few years later. Baum, who had not yet discovered that writing fiction was his true calling, spent his two-and-a-half-year stint in Dakota Territory bankrupting a variety store called Baum's Bazaar and then alienating most of the readers and advertisers of a local newspaper he edited and published. When it became clear, as Baum said later, that "the sheriff wanted the paper more than I," he and his devoted wife and children hightailed it to Chicago.

Baum's stint in Dakota was not a complete loss. When he wrote *The Wonderful Wizard of Oz* in 1899, Baum based his descriptions of the "great gray prairie" of Kansas (which he had never seen) on his memories of the landscape around Aberdeen in the hot dry summer of 1888: "Not a tree nor a house broke the broad sweep of flat country that reached the edge of the sky in all directions. The sun had baked the plowed land into a gray mass, with little cracks running through it. Even the grass was not green, for the sun had burned the tops of the long blades until they were the same gray color to be seen everywhere." Baum used the catchall prairie adjective *flat* advisedly, for this stretch of South Dakota was once the bottom of a shallow hundred-mile-long glacial lake that silted in gradually to form a broad sweep of exceptionally flat country.

That one writer of standing should have turned up in Brown

County in the 1880s is curious. That two literary lions should have stalked this sweep of prairie in the same decade seems downright bizarre, yet there was the young Hamlin Garland just a few years earlier, toiling away on his father's claim not a dozen miles north of where Baum set up shop. Or perhaps not so bizarre since this was the decade of the Dakota Boom, what Garland called the "mighty spreading and shifting" that heaved hundreds of thousands of immigrants from all over the country and the world onto the Dakota prairie, the Garlands, the Baums, and the Allens among them. The ambitious and restless, the poor and desperate, the gullible, the land hungry, the exile from oppression, the start-over dreamer, the Go West! hothead, the get-rich-quick drifter—all were spellbound by the mystique of Dakota in the 1880s. The territory's population nearly quadrupled during that decade, from 135,177 to 511,527, and the number of farms increased almost five-and-a-half-fold from 17,435 to 95,204. At every train stop, towns popped up like mushrooms after rain. "Language cannot exaggerate the rapidity with which these communities are built up," marveled one contemporary observer. "You may stand ankle deep in the short grass of the uninhabited wilderness; next month a mixed train will glide over the waste and stop at some point where the railroad has decided to locate a town. Men, women, and children will jump out of the cars, and their chattels will be tumbled out after them. From that moment the building begins." Garland, whose parents homesteaded near the town of Ordway, north of Aberdeen, in the same year that the Allens moved out to Groton, wrote that the builders of these new prairie towns labored like zealots caught in the spell of a collective delusion: "The village itself [Ordway] was hardly more than a summer camp, and yet its hearty, boastful citizens talked almost deliriously of 'corner lots' and 'boulevards' and their chantings were timed to the sound of hammers."

This was exactly the kind of carnival the Allen family got swept up in as soon as they arrived in Groton. With their piano and library and relatively comfortable house, they offered a small oasis of civilization on Main Street, and visitors flocked to them. Litigants dropped into W.C.'s office, teachers and prospective brides consulted with Edna, young shopkeepers or farmers stopped by for advice or medical care. When the beds and floor overflowed with overnight guests, Edna spread a blanket on the piano top. For the Allen boys it was bliss. The teenage Will and Hugh promptly went to work setting type for the local newspapers, and Will, who was especially enterprising, petitioned the City Council for permission to open a lemonade stand on Main Street—which his father, as police justice, denied. Young Walter, meanwhile, was collecting rattlesnake skins on the prairie (which rolled right up to the edge of their backyard), falling down badger holes, tumbling off roofs, messing about in the James River, and fooling around with guns. (When he was seven, Walter and a friend fired off a shell from a single-shot .22 rifle that grazed the forehead of one of their female neighbors, a fierce Prussian woman named Mrs. Messerschmidt, which landed them in W.C.'s "court.") And of course, for a few months every fall and winter, there was school. Groton's citizens organized a school district in February 1882, after the initial frenzy of construction died down, and by the following December, the town's first schoolhouse was finished—a rather grand two-story four-room hip-roofed frame building set all by itself on the bald, bare prairie half a mile west of Main Street. The idea behind this remote location was that the town was expected to expand and engulf the school—but meanwhile, until that happened, the Groton school stood like a sentry at the edge of the great flat abyss. W.C. naturally took a place on the Groton board of education.

Fifty pupils walked out on schooldays, cutting through the backyards of the houses and business buildings that stood one deep along Main Street, skirting outhouses and heaps of barrels and crates, dodging the fence of the St. Croix Lumber Yard, and finally ascending the imperceptible rise to what would be dubbed Presbyterian Hill. Jessie Warren, the district's first teacher, earned thirty-eight dollars a month for her troubles, which was good pay for a female teacher back then (the men usually got between 30 and 50 percent more). A few years later she was succeeded by a hapless young "professor" from an Indiana business college who had never taught school before. Somehow Walter Allen discovered that the teacher was frantically cramming every night in an attempt to stay one step ahead of the class, and he hatched a scheme to unmask him. Walter and his accomplices, who couldn't have been more than seven or eight years old, studied their "fool heads off," as family lore relates, reading pages and pages ahead of the daily assignments and easily outpacing their teacher. In class each day the boys would torment the "professor" by stumping him with questions about material he hadn't gotten to yet. This is quite sophisticated compared to the typical prairie school prank of strong-arming the teacher outside the schoolhouse and then locking the door. Being "turned out" or "carried out" by the bigger farm boys was the worst nightmare of many a country schoolteacher. One Indiana schoolteacher was only able to regain control of his schoolhouse by capping the chimney and literally smoking the children out. Or so the story went.

But Walter Allen was no rude farm boy, as he would have been the first to point out. A wry sense of humor, a streak of mischief, and the twinkling conviction that he was clever enough to get away with anything were traits he shared with his older half brother Will. The two boys, though born ten years apart, were extremely close. Will and Walter were both impulsive, and in difficult situations both preferred to take matters into their own hands rather than heed their elders.

Like a lot of brothers, they seemed to have an instinct for knowing when the other was in trouble—which in those days on the prairie almost always meant being on the wrong side of an act of God.

CHAPTER TWO

Trials

God inflicted ten plagues on the Egyptians to punish them for refusing to free the Israelites, but with the settlers of the North American prairie He limited himself to three: fire, grasshoppers, and weather. The stories that the pioneers made of their lives were essentially about how they coped with the hardships these plagues left behind.

A prairie fire swept through the Schweizer settlement just days after the families settled in Dakota. They stood on the treeless land and watched the flames travel with unbelievable speed over the dry autumn grass. Clouds of smoke blotted out the sun. The heat was unbearable. The Kaufmanns and their neighbors in Rosefield Township escaped, but others lost everything—the trunks they had hauled from the Ukraine, the lumber they had purchased in Yankton, the sod houses they had sweated to build. One pioneer boy remembered the prairie fires of his childhood as "a strange glare against the window" that would haunt his sleep on summer nights.

"Upon looking out, I saw a great wave of fire, a moving wall of flame, pass by our house and going on to the south." When the fires passed, the boy wrote, the prairie was a black expanse "dotted with ashpiles which in many cases, as though they were tombstones, marked the graves of all the settlers' material possessions."

Fire destroyed utterly and sometimes killed, but if anything, the settlers hated the swarms of grasshoppers—the now extinct Rocky Mountain locust species *Melanoplus spretus*—even more than fire because the insects were alive and conscious and seemingly perverse in their intentions. All summer long the crops would grow beautifully, filling the farmers' hearts with hope, and then on a sultry windy afternoon a mass of locusts would descend from the sky, and in hours they would strip the fields bare. "Tragic, abominable injustice," Hamlin Garland railed when grasshoppers cleaned out his parents in the early 1880s. A single swarm, according to early settlers, could be a mile high and a hundred miles across—one hundred billion bugs moving east at the rate of five miles an hour like an immense atmospheric stain. The air became so thick with insects that "the light took on a gray flickering look" according to one pioneer. "They drifted over in such clouds as to blacken the whole heavens," another prairie settler wrote of the locusts that descended again and again in the 1870s, "and with such a buzzing, roaring noise that it could be heard a long time before they came over us. . . . When they settled down the corn and vegetables would be so completely covered as to be black with them one over another. The corn was their first choice. When they had stripped it of every particle of foliage—which they would in a night—they would stick so thick on the stumps of stalks that there would be no room to stick the point of a finger. . . . As we walked along they would rise from the ground in such clouds and swarms that we had to fight our way through them. It was a time when nobody needed to be admonished to keep his mouth shut."

This is exactly what the Schweizers experienced during their first two summers in Dakota. Some potatoes and a few bushels of wheat were all Johann Kaufmann was able to salvage in the summer of 1875, and the next summer was worse. The insects waited until August of 1876, just weeks away from the grain harvest, and then descended on the fields in ravenous clouds. The day after the swarm landed, the wind shifted to the south and blew without cease for the next two weeks, effectively pinning the grasshoppers in place. By the time the cursed insects left, the crop was utterly destroyed. The loss of the second crop was devastating to the struggling families. Many considered returning to the Ukraine—but they knew it was impossible. "They had burned their bridges behind them," one of their children wrote, "and were now destined to live or die on the frontier."

The Rollags endured the same devastation on their homesteads in southwestern Minnesota. "One day we thought it was raining," recalled Gro, "but instead of drops of water rattling on the roof boards, it was grasshoppers. We looked at our little garden and potato patch and it wasn't long before everything was taken slick and clean all around us. . . . We had 60 acres of wheat sowed, but we only harvested 13 bushels more than we seeded." Gro's brother Osten told his grandchildren that after devouring "every green living thing in their path" the hoppers would attempt to gnaw the wooden handles of the farm tools. Others watched helplessly as they went after fences, curtains, furniture, clothing. After losing all their crop in the plague years of the mid-1870s, the Rollag men were forced to get jobs laying track for the Great Northern Railroad in order to earn enough money to feed their families. They hated taking orders from gang bosses, they hated being taunted by Irish workers for their Norwegian accents, they hated being away from their families and fields. Like the Schweizers, they thought about leaving but they were too poor to move.

Weather, the third of the prairie plagues, was in fact the root cause of all the other miseries. Fire, grasshoppers, bad harvests, disease, the deaths of children—whatever went wrong in their lives—ultimately came from bad weather. None of them, even the families who had relocated from other parts of the country, were accustomed to the pace and the scale of prairie weather. The ceaseless wind, the epic lightning storms, the abrupt irrevocable droughts. The sky was so immense, the atmosphere so volatile that it only heightened the monotonous absences of the earth: absence of trees, landmarks, features, variety. But when a blizzard struck, the very absence was erased. "When the fierce winds swept the blinding snow over hill and valley, everything looked alike and it was almost impossible to find your way," Norwegian immigrant Lars Stavig said of his new home in Day County, Dakota Territory. "Many a brave pioneer who came out here with great hopes and plans for a long, prosperous and happy life, in his own home with his family, was cut down in the prime of life. This cruel, treacherous enemy, the blizzard, spared no one." A blizzard sent everything visible streaming sideways before their eyes; no sound could be heard but the rush of wind and sometimes at the edge of the mind a howl rising in the distance, then lost again in the blast. In a blizzard the essential conditions of their lives—their solitude, their exposure, the distances between their houses, the featurelessness of the landscape, the difficulty of communication—turned against them. Only a few steps away from shelter, death was waiting, though plenty of settlers died inside, too, when the cold was too much for the piles of coal, twisted hay, dried animal droppings, or bones that they burned for fuel. If limitless space was the ultimate blessing of the prairie, a blizzard was the ultimate curse. It was the disaster that epitomized all the others.

And so every pioneer narrative from the prairie includes a reckoning of the worst blizzards. Rarely do they embellish or blur the facts with emotion. The assumption is that the reader will know what it feels like. But still there is the compulsion to set down the essentials—where they and family members were when the storm hit, how they got home or why they didn't, what they burned to stay alive, how long the storm lasted, when and where the victims were found. Survivors' stories.

The first bad blizzard came on January 7, 1873, and blew without cease for three days. Tilla Dahl, the daughter of Norwegian immigrants who had settled in Minnesota's Blue Earth County, remembers that her mother was out visiting neighbors when the storm struck. Tilla's father, Niels Dahl, concluded his wife was lost and decided he must go out in search of her. Before he left he filled the cookstove with wood, drew up three chairs a safe distance from the fire, and instructed his three daughters—Tilla, four, Caroline, six, and Nellie, eighteen months—to sit in the chairs, fold their hands in their laps, and repeat the Lord's Prayer until he returned. Under no circumstances were they to leave the chairs. Astonishingly, the children obeyed, and Niels found them just where he had left them when he returned safely with his frightened wife. Tilla wrote that at some point during the storm the temperature fell to 40 below zero.

Seventy people died in Minnesota during that January blizzard, some from families so poor that the bereaved could not attend the funerals because they didn't have enough clothing to venture out. The Minnesota legislature appropriated five thousand dollars for relief of storm victims, but the funds were not even sufficient to pay the doctors who cared for the frostbite victims.

Another three-day blizzard arrived two months later, in early March, after a thaw had melted some of the snow and muddied the fields. The wind came so suddenly that it sucked up mud from the fields and spat it into the blowing snow. On the Henjum farm between Wells and Blue Earth, Minnesota, drifts quickly covered the stables and shacks where the family kept their animals, and the chickens froze to death. When their fuel ran out, the Henjums stayed warm by cutting the tops off the saplings they had planted as a windbreak and feeding the green sticks into their stove. But it wasn't all grim horror. Between the granary and the pigpen the wind spun a fantastic delicate white mountain. "The snow had whirled and piled up into a mountain 62 feet high, actual measurement," recalled the Henjums' daughter. "The mountain was a beautiful sight, reaching to a thin point at its uppermost peak."

The next blizzard, which followed a few weeks later, in April, is still talked about in Yankton, South Dakota, because it came when George Armstrong Custer was quartered in town. Lieutenant Colonel Custer had been assigned to frontier duty in the Dakotas early in 1873, and he traveled west with a company of eight hundred officers and enlisted men from the Seventh Regiment of the United States Cavalry along with his devoted wife, Elizabeth, and forty government laundresses. Ill when the blizzard hit, Custer weathered the storm in the comparative comfort of a cabin attended by Elizabeth, while scores of his men wandered lost in the blast after their tents blew over. Winds at Yankton blew at an average velocity of 39 miles an hour for nearly a hundred hours, and for the entire twenty-four hours of April 15 the average wind speed exceeded 52 miles per hour. Townspeople rallied round and eventually gathered in the missing soldiers and laundresses, including one who had a newborn baby. Custer later officially commended the good people of Yankton for saving "the lives of a great number belonging to this command, besides saving the government the value of public animals amounting to many thousands of dollars." Three years later, he was dead at the fiasco of Little Bighorn.

General Adolphus W. Greely, who was head of the nation's weather forecasting service from 1887 to 1891, wrote in his 1888 book *American Weather* that "shortly after this storm the use of the word [*blizzard*] became tolerably frequent in the northwestern parts of the United States, to indicate such cold anti-cyclone storms as are attended by drifting snow."

They called the winter of 1880-81 the Snow Winter because the snowstorms started early and never let up. A three-day blizzard took the settlers of the Upper Midwest by surprise on October 15, and after that, snowstorms came at regular intervals through the winter and into the spring. In some places snow from that first October storm was still on the ground come May. Mary Paulson King, a child of immigrant Norwegian parents in Yellow Medicine County, Minnesota, remembers opening the door on the morning of October 15 to a wall of snow that "just fell in the house." Her father had to get up on a chair and make a hole in the snow in order to crawl out. After that the blizzards broke in waves—"almost one continued blizzard," according to a Dakota pioneer. Children sledded from the peaks of their roofs all winter. Soddies and one-story shacks were entirely buried in snow, but even substantial two-story homes had snow up over their second-floor windows.

No one was prepared for deep snow so early in the season, and farmers all over the region were caught with crops to harvest and fuel supplies low. Like most, Johann and Anna Kaufmann had not yet milled their grain or dug out their potatoes when the first blizzard of the Snow Winter arrived. At first they were sure that the snow would melt and there would be time to haul the grain to the mill before winter really set in; but as the weeks passed and new storms kept piling the snow higher, they realized they were trapped with no prospect of grinding the wheat harvest into flour for bread. What made it harder was now there were three children to feed. At last, after losing the three babies—one in the Ukraine, one on the voyage to America, and the third during that first bitter year in Dakota—Anna and Johann had three healthy sons. Johann, their oldest, was nine, old enough to help his father with the animals and maybe hold the plow come spring. Heinrich was three and their baby, Elias, would turn one that coming May.

By Christmas, starvation loomed again, just as it had those first two winters. Anna heard that some families were boiling their unmilled wheat kernels into a kind of mush, but she knew she could not keep her children alive on that diet. Without flour they would never survive the winter. Finally, when it was clear that the weather would not break, six Schweizer farmers decided to make the twenty-mile trip to the nearest mill together: Each farmer took a wagon loaded with grain sacks and a team of horses, and each team broke trail for half a mile or so until the animals were exhausted; then that team would drop to the rear and the next in line would break through the drifts for the next half mile. It was a long grueling trip, but the men returned with flour, and Anna was able to bake bread for her family.

The snow was so deep by January 1881 that train service was almost entirely suspended in the region. The railroads hired scores of men to dig out the tracks, but it was wasted effort. "As soon as they had finished shoveling a stretch of line," wrote Osten Rollag, "a new snowstorm arrived, filling up the line and rendering their work useless." The blizzard of February 2, "a terrible storm with thunder and lightning and very soft snow," according to Osten, halted rail traffic to Sioux Falls completely. The trains did not run again until June 15—four and a half months later. As the Snow Winter wore on, the suffering of isolated farm families became acute. Without train service there was no food to be had in towns and the deep drifts made it impossible to haul wagonloads across the prairie. Families who had neglected to get their milling done before the October 15 storm were reduced to grinding wheat in coffee mills— a tedious procedure that required almost continuous grinding to supply enough flour for a family. Many ground the seed grain intended for spring planting and lived on that, or tried to. Mary Paulson King recalled that her Norwegian parents became so desperate for coffee that they improvised a substitute called "knup." First they would cook potatoes, then mash them and mix in flour and graham. This mush was rolled out to the thickness of a piecrust, cut into tiny morsels about the size of coffee beans, and browned in the oven. Then the toasted bits were ground in the coffee mill and brewed into knup. The Rollags improvised coffee by scorching kernels of rye and wheat. "This they called coffee," wrote Osten, "but 'hu-tu-tu' what coffee!"

Civic disaster requires a hero. Minnesotans found or created one in a young storm survivor they christened "Minnesota's Frozen Son." Michael J. Dowling was fifteen when he came within an inch of freezing to death in one of the blizzards that winter. Dowling's frostbite was so advanced that he lost both legs below the knees, his left arm below the elbow, and all the fingers and most of the thumb on his right hand. But Dowling was a fighter. He lived on to became a teacher, newspaper editor, and eventually speaker of the house of the Minnesota State Legislature. "It is what one has above the shoulders that counts," he always told fellow amputees.

When the snow finally melted in the late spring of 1881, huge sections of the prairie were flooded. Children remember parents rowing boats to town over their corn and wheat fields. Most of the town of Yankton, in what is now South Dakota, was washed away when the Missouri River overflowed its banks, and downriver the town of Vermillion was also wiped out.

Laura Ingalls Wilder made the Snow Winter the subject of her novel *The Long Winter.* Every detail in the book matches up exactly with the memoirs of pioneers: the grinding of wheat in coffee mills, the endless hours of twisting prairie hay for fuel, the eerie gray twilight of the snowed-in houses, the agony of waiting and hoping that the trains would get through, the steady creep of starvation when they failed to yet again. By midwinter, Laura and her sisters had learned to scan the northwest horizon for *the cloud,* the single sooty cloud, that presaged another storm. Even the rare sunny days only heightened their anxiety. "No one knew how soon the blizzard would come again," wrote Wilder. "At any moment the cloud might rise and come faster than any horses could run."

Before the storms shut down her country school, Laura and two of her friends gathered to chat during recess. Like the prairie memoirists, Wilder was matter of fact about the vulnerability of

these children. The Little House Books were made into a syrupy
television series in the 1970s and 1980s, but the books themselves
are spare and unsentimental. Wilder took it for granted that
schoolgirls, don't flinch when the conversation turns to death by
exposure:

"What would you do if you were caught in a blizzard,
Mary?" Minnie Johnson was asking.

"I guess I would just keep on walking. You wouldn't freeze
if you kept on walking," Mary answered.

"But you'd tire yourself out. You'd get so tired you'd die,"
said Minnie.

"Well, what would you do?" Mary Power asked her.

"I'd dig into a snowbank and let the snow cover me up. I
don't think you'd freeze to death in a snowbank. Would you,
Laura?"

"I don't know," Laura said.

"Well, what would you do, Laura, if you got caught in a
blizzard?" Minnie insisted.

"I wouldn't get caught," Laura answered.

Only the worst winters got named. After the Snow Winter of
1880–81, the next one worthy of christening was the winter of
1886–87: the Winter of Blue Snow. A beautiful name for a terrible
season. It was the winter that killed the cattle kingdom that had
flourished for nearly a decade on the western prairie. Or rather finished it off, for the seeds of devastation were sown the summer before. The weather turned unusually hot and dry early that year, and
by the Fourth of July the grass was parched and brown and stubby.
The young Teddy Roosevelt, traveling through the north part of
Dakota Territory on the way to his ranches near Medora, told a
newspaper reporter in mid-July that "Between the drouth, the

grasshoppers, and the late frosts, ice forming as late as June 10, there is not a green thing in all the region." The drought was bad in the south as well, so all summer long, more and more cattle were shipped or herded from Texas and Kansas and Oklahoma onto the northern plains. Dakota and Montana were seriously overgrazed when the summer fires started.

Even by American standards, the transformation of the short-grass western prairie from Native American buffalo hunting ground to an internationally financed beef industry had happened extraordinarily quickly. The Teton Sioux tribes retreated to Wyoming reservations in 1878, and immediately afterward Texas ranching operations began expanding northward, running the no-toriously rugged and ornery Texas longhorns by the thousands onto the plains of Montana, Wyoming, Nebraska, and the Dakotas. The "empire of grass" was the largest unfenced pasture in the world— free, seemingly endless, unbelievably productive. And potentially hugely profitable. Foreign investors, many of them "swells" from titled families in England and Scotland and blue-blooded merchant princes from the East Coast cities, pumped in millions. Books like General James S. Brisbin's *The Beef Bonanza: or How to Get Rich on the Plains* promised staggering surefire returns. The number of cat-tle run on the plains increased exponentially, especially once the buffalo herds were reduced to bands of stragglers in the mid-1880s. Suddenly ranchers had names like Teddy Roosevelt, Randolph Churchill, and Antoine de Vallombrosa, the Marquis de Mores. Polo ponies were stabled next to working ranch horses; castles and chateaux rose on the range; chefs and valets were imported from Europe. The biggest spreads were larger than Eastern states. Every-thing in the Cattle Kingdom was on an enormous scale—the land, the herds, the flow of capital, the dreams, the disasters. When the inevitable bust came, it was an epic bust.

The prairie fires of the hot, dry summer of 1886 lit the fuse. Huge swaths of short brittle grass went up in flames during July and August. "The fires have devastated a large amount of grazing," reported an eastern Montana newspaper on August 1, "and, as is usual, the very best of that." Going into the winter, the cattle were stressed and stringy when they should have been sleek and fat. Even a mild winter would have taken a toll, but the winter of 1886-87 was not a kind one. The bitter weather started before Thanksgiving. A storm blew through on November 22 and continued for days. Then, in early December, temperatures moderated for almost two weeks, thawing the drifts to a heavy slush. But when this warm spell ended, the temperature did not rise above freezing again for months. An impenetrable crust formed on top of the refrozen slush. Cattle desperate for food cut their muzzles on the shards of ice that covered the sparse grass. Steers bled to death when the crust gave way beneath them and the ice sliced open their legs. By January, the winter storms were coming in earnest. Every five days or so, the cycle repeated: two days of blizzard, three days of glittering blowing chill, then a few hours of smoky calm, and then another blizzard.

The worst storm came on January 28, 1887, with seventy-two hours of fiercely blowing snow and arctic temperatures. The storm left millions of cattle dying or dead on the range. Cattle had drifted hundreds of miles before they froze to death or died of exhaustion or suffocated from the ice plugging their nostrils. Some herds were never found; some were found in riverbeds or ravines, heaped up like slag; some were so badly frostbitten that ranchers were reduced to salvaging their hides.

Come spring, when the snow finally melted, flooded rivers carried the carcasses of thousands of cattle that had frozen to death during the winter—raging torrents choked with dead animals wedged between ice floes. Teddy Roosevelt's ranch foreman, Bill Merrifield, reported that "the first day I rode out, I never saw a live animal." Contemporary reports put the death toll at ten to twelve million head of cattle—losses of 80 percent in some regions.

"About seventy-five percent of the 'she' stuff died," rancher Charles H. Rowe from Mandan, North Dakota, wrote in his diary. "Anything that would live through that winter would live through Hell." "Well, we have had a perfect smashup all through the cattle country of the northwest," Roosevelt wrote to his friend Henry Cabot Lodge when he finally traveled out to Dakota to assess the situation for himself. "The losses are crippling. For the first time I have been utterly unable to enjoy a visit to my ranch."

The relentless blizzards and freezes of the Winter of Blue Snow proved that the "open range" system—running cattle on the western prairie without supplemental food or winter shelter—was just another foolish American dream, like the fantasy that rain follows the plow. Scores of the big ranches went under. Blue-blooded investors pulled out or declared bankruptcy, and lots of ordinary ranchers with a few dozen head were clamoring to sell, cut, and run. Roosevelt hung on for a few years, though with much smaller herds. He never recouped his initial investment of $85,000.

❧ ❧

The next winter, the winter of 1887–88, was another bad one, especially in the wheat and rye and potato country east of the 100th meridian where families like the Kaufmanns and the Rollags were finally beginning to get on their feet. "Two months of zero weather," was the way H. G. Purcell, a schoolboy in the eastern part of Dakota Territory, remembered the start of that winter. In Jerauld County, farther west, the storms began in November and intensified as the winter advanced, with the snow getting deeper and the cold more intense, week after week, through December and early January. George W. Kingsbury, author of the first comprehensive history of the Dakotas, published in 1915, pronounced that winter "unusually rigorous," with frequent heavy snows that "blockaded" the railroads and drifts that "rendered the [wagon roads] almost impassable." "There must have been considerable suffering in the newer settlements," wrote Kingsbury, "where the recent settlers had not prepared for a season of such extreme severity, and these newer settlements were in large number for Dakota had gained many thousands in population during the season preceding." Kingsbury need not have been so tentative. Nobody, not even the legendary oldest settler, was ever really prepared for the extreme severity of these seasons.

The coldest weather came, as it usually does, in December and hung on through January. Just before the New Year, a flow of arctic air pooled over the Upper Midwest and settled in like fog in a river valley. The Minnesota State Weather observer at Pine River Dam recorded a minimum temperature of 46 below on December 29; observers at Pokegama Falls and Leech Lake Dam were unable to take temperature readings that day because the mercury inside their government-issued thermometers froze solid. It's hard to find vocabulary for weather this cold. The senses become first sharp and then dulled. Objects etch themselves with hyperclarity on the dense air, but it's hard to keep your eyes open to look at them steadily. When you first step outside from a heated space, the blast of 46-below-zero air clears the mind like a ringing slap. After a breath or two, ice builds up on the hairs lining your nasal passages and the clear film bathing your eyeballs thickens. If the wind is calm and if your body, head, and hands are covered, you feel preternaturally alert and focused. At first. A dozen paces from the door, your throat begins to feel raw, your lips dry and crack, tears sting the corners of your eyes. The cold becomes at once a knife and, paradoxically, a flame, cutting and scorching exposed skin.

Temperatures in the prairie states fell even lower in January. A run of record-breaking cold commenced on the eighth and continued but for a single day's hiatus until the twenty-second. "This is an exceptionally long period for such extreme cold weather to prevail even in this climate," noted William W. Payne, the Carleton College astronomy professor who founded and ran the Minnesota

State Weather Service, "and at most stations it is unparalleled by any previous record."

On Wednesday, January 11, the citizens of Aberdeen in Dakota Territory marveled at the beautiful display of sun dogs blazing in the pale blue sky—three brightly colored spots following the sun. Though beautiful to behold, the longtime residents knew that this did not augur anything good. Local lore had it that sun dogs are a sign of approaching cold: the brighter the dogs, the colder the weather. Supposedly, when Indians saw sun dogs in the sky, they piled more wood on the fire and sealed the skin tight over the tepee door.

It was 20 below zero at dawn in most of Dakota that morning. How much colder could it get?

Two solid weeks of iron weather—except for one day, January 12. "That morning was the most beautiful morning I had ever seen," wrote Josephine Buchmillar Leber, the daughter of German immigrants, recalling that Thursday morning in Turner County, Dakota, when she was twelve years old and begging her father to let her set out for school. "Sun shone bright. It had snowed the night before. The snow flakes layed [*sic*] loosely on drifts, just like loose feathers, and as I remember it seemed the sun shining on the snow, caused a golden reflection on the snow." Another settler remarked on the "almost mysteriously velvety" quality of the air that morning. People came out of their houses and sod huts to gaze, blinking in awe, at the eerie "copper" color of the morning sky. Thomas Pirnie, a youth in Buffalo County in central South Dakota, remembered that when he awoke at daybreak on the twelfth "the air was like that of an April morning, with just a breath of breeze coming out of the southwest. I happened to be the first one of our family to go out. I quickly returned inside and called out so all could hear me, 'Oh come folks and see what a beautiful morning it is. It is 32 above. We're going to have a January thaw.' Cousin Hugh and myself took a shovel and a pan of chicken feed to the barn, expecting to soon dig our way into the sod barn of which only the roof pole were visible above the great snowdrifts that almost filled the deep ravine."

Everyone who wrote about January 12 noticed something different about the quality of that morning—the strange color and texture of the sky, the preternatural balminess, the haze, the fog, the softness of the south wind, the thrilling smell of thaw, the "great waves" of snow on the prairie that gleamed in the winter sun. The one aspect they all agreed on was the sudden, welcome rise of temperature. Even allowing for the distortions of memory, which are especially acute with weather, there is an urgency and vividness to these accounts. These are the mental snapshots of the *moments before*—the last kind hours. Historian N. J. Dunham lingers over these hours of deceptive mildness in his *History of Jerauld County*.

On the morning of Thursday, the 12th of January, the wind had fallen and become quite warm. The snow was melting a little. Great banks of fog fifteen to twenty miles wide rested across the prairies from the vicinity of the Black Hills eastward into Minnesota. Between these banks of fog were stretches of country from thirty to forty miles in width where the sun shone brightly. One of these fog banks ran east and west along the C & N. W. Ry. [Chicago and North-Western Railway], through the central part of Beadle, Hand and Hyde counties. Over all of Jerauld county the morning was warm and bright.

Farmers took advantage of the pleasant weather to go to town or to go to fetch hay from the prairie. All felt a relief from the rigorous wintry weather that had preceded. In Jerauld county at that time were 1025 children of school age.

Owing to the balmy conditions of the air, probably a greater percentage of those children went to school that day than on any previous day for weeks.

All through the region, from the Black Hills to the homesteads of eastern Minnesota, a kind of undeclared civic holiday was being celebrated—a celebration of ordinary daily life. From nearly every home, someone set out for town on foot or by horse to replenish supplies. Farmers turned their animals out of their barns to stretch their legs, drink at watering holes, browse the piles of hay the farmers had forked together in the fall. For the first time in weeks people could be outside without being in pain. That was reason enough to celebrate. Many were convinced it was the January thaw—the start of a week or more of mild weather—though a few weatherwise old-timers and farmers with a sixth sense about the atmosphere sniffed the suddenly balmy air suspiciously. Almost a 40-degree rise in twenty-four hours—it didn't sit right somehow. Certainly not to John Buchmillar, who decided to keep his daughter Josephine home from school that morning despite her hot tears and wails of protest. "I feel there is something in the air," Josephine overheard her father tell a neighbor solemnly at about eleven o'clock that morning.

Maria Albrecht also had a bad feeling that morning. The day dawned dull and cloudy over the Schweizer farms in Rosefield Township, and there was fresh snow on the ground from snow showers that had blown through the previous day. Like everyone else in the region, Maria noticed the unaccustomed mildness—but something about the look of the sky bothered her. She couldn't name it or explain it, but there it was. From the moment she had gotten up she could barely keep from crying. And so, while her husband Johann was out in the barn tending to the animals and her children were getting their breakfast in the dim morning light, Maria made up her mind. The boys would not be going to school that day. And anyway, it was her husband's forty-first birthday. That was reason enough to keep the boys home.

Of all the neighborhood families, the Albrechts lived farthest from the one-room schoolhouse—the English school, as they called it—that the Schweizers had built a few years ago at the "middle fence" (the midpoint) of the western side of section 26. For the Kaufmann boys the walk was nothing—half a mile and they were there. And the Grabers with their crowd of children practically had the school in their yard. Seven children Peter and Susanna Graber had had since they married in 1875—with seven others from Peter's first marriage. Fourteen young Grabers, while Maria and Johann only had five—Johann, the boy who had been born on board the *City of Richmond,* who was now thirteen; nine-year-old Peter; Anna, six; and two little ones, Jacob and Julius. Even Anna Kaufmann, who had endured so many losses, now had more children than Maria—six living children. Her Johann still attended school, even though he was a young man of sixteen. His brother Heinrich, ten, and Elias, seven, walked to school with him when the weather wasn't too bad. And there were three other Kaufmanns still at home—six-year-old Julius, three-year-old Jonathan, and Emma, who had celebrated her first birthday on New Year's Eve 1887, just twelve days before.

The Albrechts, with fewer children than the other families, had agreed to house and feed the schoolteacher, Mr. James P. Cotton. It was a difficult arrangement because Mr. Cotton spoke no German and the Albrecht parents spoke little English. Maria put up with it because having the teacher around made her feel a little easier about sending the boys off across the prairie. But not that morning. This was not a day for them to leave the house, she was sure of it. There was a quarrel as the two older boys were getting ready for school. Maria insisted they stay home. Johann insisted just as adamantly that they must attend—Mr. Cotton had expressly told them not to miss that day. Maria called her husband in from the barn to lay down the law. But Johann Sr., who was used to these disputes, took the boys' side. What would it hurt if they went to school on such a warm morning if that's what they wanted to do? Johann shrugged off his wife's pleas and went back to work. The fact that it was his forty-first birthday would not stop him from putting in a full day of work, especially on such a promising winter day.

So thirteen-year-old Johann set off across the field alone. Peter, to please his mother, stayed home, though it made him miserable to miss a day of school. Years later, Peter recalled that he spent the morning sitting by the window staring glumly in the direction of the school. "My mother, who was observing me, sighed and said, 'My child, my heart is going to break. I wish your brother were here with you.' My father was busy working outside."

The Allens were asleep in their house on Main Street when the weather changed, so they never knew at what hour the winter westerlies flickered and finally died out altogether and the south wind rose in their stead. When they stirred on the morning of the twelfth, the temperature in Groton was pushing 20 degrees—positively balmy compared to the 20-below-zero reading of the previous day. You notice a rise in temperature like that when you wake up in the dark in a house with no fire. But the three Allen boys—the teenagers, William and Hugh, and their eight-year-old half brother, Walter—were comparatively lucky. Being town boys, they didn't have to go out to the barn to feed the stock or chop through the ice on a frozen trough before they got their own breakfast. The boys could get up and dress and eat breakfast in the daylight. And for once it looked like a nice day.

Unlike some boys, who had to be prodded and badgered out the door every morning, Walter Allen was always eager to get to school because he had a special job to do. In the Groton school, each row of desks was under the supervision of a "row monitor" who was in charge of the coats and overshoes for all the children in that row. As monitor of his row, Walter commanded the front seat, and whenever school was dismissed or recess called, he got to jump up before the other children, rush to the vestibule where the coats, caps, scarves, mittens, and overshoes were stowed, gather them up, and distribute them to the children in his row. (This strict segregation of children and clothing was enforced in order to keep the odor of wet wool, felt, and leather out of the classroom.) Walter was extremely proud that he had already mastered the tricky business of matching kids and clothes and even mentioned this accomplishment in his diary.

So that Thursday morning, while W.C. Allen ducked out the door and went to his law office next door to attend to the multifarious affairs of running a boom town on the Dakota prairie, and W.C.'s two older sons, Hugh and Will, strolled down main street to their jobs at the Groton newspaper (reduced to one now), Walter Allen set out alone for school with all the brisk determination that an ambitious eight-year-old can muster.

❧ ☙

At some point during their first few years in Minnesota, the Rollags had quietly altered their names to make them sound more American. Gro became Grace, Osten became Austin, Gro and Ole's oldest child, Peder, born in November 1874, just five months after they claimed their homestead, became Peter. Carl, Grace and Ole's second son, born two and a half years later in the same sod house as Peter, became Charley. Though the boys went to English school and quickly learned to speak English with barely a trace of Norwegian accent, they continued to speak Norwegian at home. Thanks to their grandmother, Kari, they also had plenty of Norwegian books to read—indeed, Kari saw to it that they read Norwegian even before they started at the English school, which she didn't think very highly of anyway. Kari also insisted that the children be confirmed in Norwegian, since she was dubious whether being confirmed in English would really "take." Kari was one of those grandmothers who leave an indelible impression on her grandchildren. On winter evenings she mesmerized the boys with stories of the first years on the prairie when green-eyed wolves trailed after her whenever she ventured out after dark and Indians turned up at her door begging for food and snakes slithered out of the sod house walls in the early spring, coiling themselves on the roof and basking in the sun. Grace and Ole had finally replaced their soddie with a frame house after the Snow Winter. It had taken them seven years.

Kari was sixty-five years old in the winter of 1887-88, her daughter Grace, thirty-six. Grace and Ole had had seven children by then, though only six were alive since they had lost a daughter, Anna Marie, at fourteen months in October 1885. Peter and Charley, the two oldest, were now thirteen and ten, both handsome boys, Peter blond and thin with close-set eyes, a narrow face, and long, delicate fingers for a farm boy, Charley dark-haired and clean-featured and bigger-boned than his brother.

For some reason the boys did not go to school on the morning of January 12. Forty years later, when she wrote her "Recollections from the Old Days" in Norwegian, Grace recorded everything about that day except why the boys were not in school. Possibly the English school had already closed for the season. Or possibly Grace and Ole kept the boys home to work on the farm since it was the first day in weeks that they could get the cattle out to the springs and bring in the hay they had cut and stacked in the fall. By mid-day both parents and the two boys had been working outside for hours. Working fast. For Grace and Ole had been in this country long enough now to know that such weather would not last long. Not in January.

The Shattucks lived in Holt County, Nebraska, for two years after leaving Seward behind, but they never really got used to it. For one thing, the people were different—Irish most of them, with names like O'Hara and Murphy and O'Donnel. A wild-eyed dreamer named John J. O'Neill—a captain of an African-American infantry unit in the Civil War and later the ringleader of the doomed Fenian invasion of Canada—had planted an Irish Catholic colony here on the Protestant prairie and named the principal town for himself: O'Neill, Nebraska. Even Emmet, the little railroad town north of Etta Shattuck's school, was named for an Irishman, bold Robert Emmet, the darling of Ireland, an early martyr in the struggle against England. The soil was different in Holt County, too—or what passed for soil. Sandy and porous, it hardly held what little rain fell. The prairie grass was shorter than that in Seward, the farms farther apart and poor and wind-bitten. Brown most of the year except in a good spring—brown and crusted white most of the winter.

Considering the thin soil and the sparse summer rain, it's not surprising that Ben Shattuck failed to make a go of his Holt County farm. The second summer he didn't even plant a crop. The only money the family could look forward to was the twenty-five dollars a month that Etta brought in from teaching school, and that wasn't enough to support the seven of them. So before the cold weather set in, Ben and Sarah Shattuck packed up and moved with their younger children back to Seward.

Etta stayed on alone to teach her country school, district 141, the Bright Hope school district. She boarded with a local family and walked to the schoolhouse every morning. Etta could stand outside the schoolhouse and count the features on one hand. A couple of sod houses and barns silhouetted against the low horizon; a rounded pile of dry prairie grass by each house; and way off to the east the bare limbs of willows and cottonwoods outlining the bed of the Elkhorn River. Between school and the river, the frozen grass and snow extended flat as paper; to the west the ground undulated just perceptibly as it rose to meet the Sand Hills. That was the Bright Hope school district.

Usually seven or eight scholars showed up, never more than ten, most with Irish names. The families were large, but only one child or two from each family attended school because that's how many pairs of shoes they had. Their parents referred to it as the Maring school because a farmer named Maring owned the land it was on—owned the entire section, all 640 acres. Ten acres or so he had in corn, another ten in oats for the livestock, a few acres of sorghum so they could make syrup and have something sweet to put on their corn bread. The rest was unbroken grazing land. Dreams of growing grain, maybe even a garden of vegetables, died hard and slowly in these parts, and Maring always talked wistfully about putting in more corn. But in most years there just wasn't enough rain. Hay was the big thing in Holt County. They cut it in July or August and loaded it in the train cars at Emmet—in fact the native hay was the reason the Elkhorn Valley Branch of the Chicago and North-Western Railway stopped at Emmet. Prairie coal or poor man's coal, they called it, and in the absence of real coal or wood or buffalo bones, it kept them warm through the winter, though the farmers and their wives and children cursed the endless task of twisting it and feeding it, hank after hank, into the hay burners on top of their stoves.

Ten days into January, Etta decided to close school. The weather had turned so frigid that the parents worried their children would get frostbite or chilblains or even freeze to death on their way to school. Families couldn't afford to lose a child to injury or death: Out on the frontier, working children made the difference between surviving and going under. Keeping children safe was more important than educating them. So Etta let it be known that the term was over. The school would open again in spring, but Etta wouldn't be there to teach it. She was going back to live with her family in Seward. She, too, had had enough of Holt County.

Before she left, Etta had one piece of business to attend to. In order to get paid for her last month of teaching, she had to have an order signed by the school district superintendent, which she would then present to the district treasurer. It seems absurd that in a district with ten students she had to go through the formality of obtaining a signed paper attesting that she had fulfilled her contractual agreement to teach "in a faithful and efficient manner . . . to keep the school-house in good repair, to provide the necessary fuel and supplies," etc.—but rules were rules, even out on the frozen prairie. Without a signed order, the treasurer could not pay her.

So on the morning of January 12, 1888, Etta Shattuck set out on foot for the house of J. M. Parkins to get the order signed. The next day, Friday the thirteenth, with her wages in her pocketbook, she would walk to O'Neill and get on a train and go back to her parents in Seward.

Etta, like her father, was a devout Methodist. Her conversion experience had come at the age of seventeen, just before the Shattucks left Seward, and her pastor at the Seward Methodist Episcopal church affirmed that her faith was "sublime." When she was alone, Etta was in the habit of singing hymns and praying silently or even out loud when the spirit moved her, so it's likely that she was lost in song or prayer as she walked out to J. M. Parkins's house that warm, breezy morning.

Later, when the newspapers scrounged for every crumb they could find, it was reported that the father of the family with whom Etta boarded shouted after her when the wind suddenly shifted and the dark cloud raced out of the northwest. The dark cloud that heralded the blizzard. He shouted for a long time, he said, hoping against hope that Etta would hear him and turn around. He shouted as loud as he could into the rising wind and the suddenly seething air, but he didn't dare venture out after her.

CHAPTER THREE

Disturbance

Though the convulsions of the atmosphere are often complex and multifaceted, extreme cold has a fairly simple formula. Diminish the duration and intensity of sunlight, deflect winds carrying milder currents, level the surface of the land, wait long enough—and you're sure to end up with a pool of dense, calm, frigid air. Just to be sure, add a layer of clean white snow so that what light does reach the ground is reflected back into the atmosphere before the earth can absorb its warmth. If, on top of all this, you have clear skies above you, temperatures will plunge spectacularly at night as the atmosphere's infrared energy radiates off into space. Of course, there are other recipes for outbreaks of cold weather, including the movement of fronts and pressure systems, but this is the easiest and surest. A week or so of these conditions and the cold will be fierce, unyielding, and deadly.

All of these conditions came into alignment over the Canadian interior during the first days of January 1888. Take out a map of Canada and run your finger along the 60th parallel—the line that runs from the Gulf of Alaska to Hudson Bay, neatly separating the tops of British Columbia, Alberta, Saskatchewan, and Manitoba from the bottoms of the Yukon and the Northwest Territories. This is one of the world's perfect breeding grounds for cold. At this latitude on the day of the winter solstice, the sun remains above the horizon for a total of 5.6 hours. For the other 18.4 hours of the day it's either pitch dark or an eerie purplish twilight, depending on the state of the atmosphere. But calling those 5.6 hours "sunshine" is a bit misleading. Even if it rises into a perfectly clear sky, the winter solstice sun over Fort Smith, Fort Resolution, Fort Simpson, Fort Liard, and Watson Lake provides essentially no solar energy. There is light, but because the rays are so slanted in winter this far north, very little solar radiation is absorbed at the surface to heat the ground. Fort Simpson, which sits on the Mackenzie River in the Northwest Territories hundreds of miles and two mountain ranges away from the moderating influence of the Pacific Ocean, has an average daily temperature of 11 degrees below zero in December and slightly over 16 below in January.

But at the start of January 1888, it was considerably colder than the average. With high pressure bearing down on western Canada, surface winds were light. Nothing disturbed the vast shallow pool of cold air that settled over the snow-covered plains and lakes. The longer the atmosphere stagnated, the colder it became. On January 3 the temperature hit 35 below zero east of Fort Simpson. Gradually, over the next few days, the cold air mass expanded and flowed southward like a glacier of sluggish gas. By Sunday, January 8, the lobes of cold had pushed as far south as Medicine Hat in Alberta, about 70 miles from the Montana border, and as far east as Qu'Appelle, due north of the Montana/North Dakota line. And there it sat, a pool of dry, stagnant, and exceedingly cold air, too heavy to rise into the warmer air above it, too inert to mix with the milder air masses around it. Imagine a blob of invisible subzero mercury sealed and quivering over a quarter of a continent.

Stable is the word meteorologists use for an air mass of this sort—but nothing in the atmosphere is stable forever. A minor shift in the flow miles above the frozen surface, if conditions were ripe, would be enough to shatter the cold's fierce grip. But it wasn't going to go quietly.

Constantly and futilely, the earth's atmosphere seeks to achieve equilibrium. Weather is the turbulent means to this perfect, hopeless end. Contrasting temperatures try to balance out to one uniform temperature, pressure differences strive for resolution, winds blow in a vain attempt to finally calm down global tensions. All of this is enormously complicated by the ceaseless rotation of the planet. Weather is the steam the atmosphere lets off as it heaves itself again and again into a more comfortable position. Weather keeps happening because the equilibrium of the atmosphere keeps getting messed up.

It doesn't help that the planet itself is irregular, with crumpled solid chunks of land randomly interrupting the smooth liquid surface of the oceans. Equilibrium doesn't stand a chance against all these complex interacting variables. There's so much going on out there—and up there—that the very striving for equilibrium is erratic, chaotic. There are patterns, of course, repetitions and cycles, long stretches of monotony and eerie symmetries, but weather, by its very nature, lacks a fixed overall structure. It's a stream that perpetually remakes the channel of its flow.

Shortly after Christmas 1887, a ripple developed in the flow about six miles above the surface that would in time dislodge the frigid air massing over interior Canada. Today, accustomed as we are to the patter of televised weather forecasts, it's easy to reduce the violent and deadly blizzard that resulted from this disturbance to a canned meteorological scenario—the low dropping down from Canada and

tracking south and east across the Upper Midwest before swinging
to the north, the blast of wind and snow that accompanied the passage of the cold front, the outbreak of arctic air that surged as far
south as Texas. But in a way this description only trivializes what
was really happening in the air and on the ground. A storm, any
storm, involves the entire atmosphere. Meteorologists refer to the
atmosphere as a "continuous fluid" or, in the words of chaos theo-
rist Edward N. Lorenz, a "thermally driven rotating fluid system,"
and the phrases are apt. In weather everything connects.

Connects first of all to the sun. The uneven distribution of the
sun's radiation is what causes temperature differences in the first
place. As areas of higher and lower pressure develop in response to
these temperature differences, winds begin to kick up, blowing
from high pressure to low pressure (an extreme example of this is a
pressurized container like a can of hair spray: Think of how the
small parcel of high-pressure air rushes out into the lower-pressure
air around it when the valve is released). The ripple that shoved the
cold air out of Canada was born of the interaction between air
masses of contrasting temperature and pressure, and so, in a larger
sense, was the upper-level flow of air that swept up the ripple and
carried it down the spine of the continent. That upper-level flow,
commonly known as the polar jet, circles the globe from west to
east at altitudes of between six and nine miles. The jet stream is a
natural boundary marker, an atmospheric river flowing between
the region of warm subtropical air to the south and cold polar air to
the north. The sharper the temperature contrast between those two
regions, the faster the river flows. In winter, the course of the river
drops south and its current stiffens—winter speeds of 75 mph are
common, though it has been clocked at 200 mph. By January the
vigorous winter jet is down around the 50th parallel—right over
America's northern tier of states.

Or rather it *would* be over the northern tier if it flowed in a straight
line, but in fact the jet meanders around the world in a series of im-

mense loops, each of which spans some three thousand or so miles. These loops, or long waves, as meteorologists call them, form as the jet gets squeezed or stretched by the prominent irregularities on the earth's surface, like major mountain ranges or the ocean basins. Typically, three to five long waves are slowly moving and evolving around the earth at any given moment. As these waves interact with the jet stream that flows through them, great eddies of air known as cyclones (lows) and anticyclones (highs) spin up, which themselves warp the whole pattern of waves. High pressure at the surface, which is associated with warm air aloft, creates a poleward bulge in the flow called a ridge—an arc that slopes to the north on its westward flank, peaks at the top of the high, and then descends southward; similarly, low pressure at the surface, associated with cold air aloft, sends the jet plunging south in an arcing trough that bottoms out around the base of the low. When forecasters speak of a "ridge of high pressure" or a "trough of low pressure," what they're talking about are the peaks and valleys of the long waves that distort the west-to-east progress of the upper flow. Each long wave is measured by its wavelength (the distance from ridge to ridge and trough to trough) and its amplitude (how far north or south the loops are deflected).

During the first week of January 1888, as Etta Shattuck was teaching her last day of school in the Bright Hope school district and Anna Kaufmann's three older sons were dutifully walking back and forth over the hard-packed snow to Mr. Cotton's schoolhouse and eight-year-old Walter Allen was studying his fool head off in Groton, a dome of intense high pressure in western North America buckled the polar jet into a high-amplitude ridge. This pattern might have hung in for a week or more before the long wave drifted away or gradually weakened, and we never would have heard a thing about it. The reason it didn't happen this way is that a sudden shot of energy surged into a segment of the jet and caused the core of the current to accelerate rapidly—like a bullet train rocketing through a tunnel of air, a tunnel that is moving as well in the same direction, only more slowly. This atmospheric bullet train is called a jet maximum or jet streak. It's hard to pinpoint exactly how and when a particular jet streak takes off, though there are several scenarios typically at play. Six miles above the Arctic, vortices of cold air spin around columns of even colder air—picture an atmospheric whirlpool six hundred miles across—and it's possible that one of these vortices, known as cold-core lows, drifted south until it collided with the ridge in the jet stream and then unraveled its energy into a jet streak. Or a disturbance off the east coast of Asia born of the contrast between mild maritime air over the Sea of Japan and cold continental air blowing off the deserts of Mongolia might have been the jet streak's energy source. From January 5 through 8, the observer at the U.S. Army Signal Corps station at Anvik in west-central Alaska recorded high northeast winds and snow, which may well have been a sign of the jet streak careening in from the Pacific. Or the band of enhanced winds might have risen out of the ghost of a storm that had burst and dissipated over Europe or central Asia days earlier and then fed its ephemeral remains into the jet.

In any case, some disturbance created a crimp, or "short wave," in the smooth undulation of a long wave, and at the heart of this crimp a jet streak spurted forward. As long as it was over the Pacific or sailing up the west side of the North American ridge, the jet streak existed as pure potential—the tightening coils of an ineffable spring. Its potential would be released only if and when the jet streak encountered the right conditions to reinvigorate it.

A week into January, those conditions presented themselves one after another in quick succession like a run of losing poker hands. There was no reason why one bad hand had to follow another—the deck was shuffled and dealt anew each time. It was just the luck of the cards—loss after loss that finally compounded into catastrophe.

On Tuesday, January 10, the jet stream with its embedded jet streak, having crested the ridge somewhere up in the northern reaches of British Columbia or the southern reaches of the Yukon, began diving southeast into western Alberta. It was here that the flow encountered the immense irregular wall of the Rocky Mountains—the first losing hand. The mountains squeezed and deflected the current, altering its temperature and pressure. As the flow descended the eastern flank of the Canadian Rockies, the air warmed, and as it warmed it dropped the air pressure at the surface. Some disturbances, known as leeside lows because they form on the eastern or leeward side of the Rockies, spring to life in this way. But given the intensity of the storm that followed, it seems more likely that the remnants of an older disturbance had come in from the Pacific and amplified when the jet crossed the Rockies. As the upper flow crested the jagged obstacle of the mountain range and soared over the great flat expanses of the North American plains, it sent an immense vortex spinning counterclockwise all the way down to the surface. The ghost had come back to life.

Propelled by the high pressure building in behind it, the low worked its way southeast down the tapering lower half of Alberta on Tuesday, intensifying as it moved. The air was so cold that it had very limited capacity to hold moisture, so not much snow fell. That would come later, when the vortex fastened onto a stream of moist air coming up from the south. The stronger the low became, the more surface air it pulled toward the center of its vortex. You'd think that the low would eventually "fill" by pulling in enough air to raise its pressure and thus fizzle itself out. The reason this didn't happen was because of the way the jet streak was roiling the flow six miles up. To continue with the bullet train and tunnel analogy: Air moving through the jet stream's tunnel was forced to converge as it got sucked into the rear, or "entrance region," of the jet streak's bullet train; but when the air was hurtled out the nose, or "exit region," of the train, it spread or diverged from the core of the

flow like a delta at the mouth of a high-speed river of air. The diverging flow aloft acted like a pump, which evacuated the air below
it. The winds converging at the surface got sucked up into the vortex of the low and then, due to the force of the jet streak, spewed
out the top by the diverging upper-level winds. With a greater
mass of air streaming out the top of the funnel than feeding into
the bottom, the air pressure at the surface kept dropping. Meteorologists call this upper air support for a developing storm.

The chance alignment of the low and the jet streak's exit region
was the second losing hand. Not only did the jet amplify the low,
but it forced it to take a steady course to the southeast. With the jet
feeding and steering it, the disturbance was cranking up into a
powerful and fast-moving low pressure system—a "mid-latitude
cyclone" in meteorological parlance.

Sometime during the first hours of Wednesday, January 11, the
advancing low crossed the U.S. border and began to cause the air
pressure over northeastern Montana to fall. All that day it continued to churn southward, until by nightfall on Wednesday there was
a well-defined trough of low pressure radiating out from the vicinity of Fort Keogh near Miles City, Montana, and extending north
into southern Alberta and British Columbia and south to Colorado.
By itself, the strengthening low would have kicked up some stiff
wind on the Great Plains, blown the snow already on the ground
into drifts, maybe spat out a few inches of new snow before subsid-
ing: a typical midwinter storm; certainly nothing historic. But as it
dug deeper into U.S. territory, the low uncovered a source of highly
explosive fuel that boosted its power exponentially.

To the north of the low, up in central and northern Alberta, that
pool of arctic air had hardly budged for a week now, and the longer
it stagnated the colder it got. To the south, a mass of unseasonably
mild and humid air from the Gulf of Mexico was beginning to
stream up over Texas and Oklahoma. The potential energy in the
temperature differential between these two sharply contrasting air

masses was enormous. In order for that potential energy to be converted into the kinetic energy of violent weather, something had to bring the air masses together—the more sudden their encounter, the more violent the weather would be. That something was the intensifying low. The fact that the low happened to wander down between these two air masses at this particular moment in time was the third bad hand. The hand that finally and abruptly ended the game.

CHAPTER FOUR

Indications

The person charged with the job of predicting the origin and movements of this spiraling atmospheric disturbance was a thirty-nine-year-old career officer by the name of Thomas Mayhew Woodruff, first lieutenant, Fifth Infantry, United States Army. Woodruff was a good man, well educated, gently born, unfailingly courteous, who took his work seriously and did his duty conscientiously. The fact that so many people died when the potential energy of this disturbance was released over the Dakotas, Nebraska, Minnesota, and Iowa on the afternoon of January 12 was by no means Woodruff's fault. Given the state of the art of weather forecasting in 1888, Lieutenant Woodruff did the best he could. He simply didn't know enough to do any better—and he didn't have the means to make effective use of what knowledge he had. It's questionable whether anyone in 1888 could have done more.

Lieutenant Woodruff's failure, if one can speak of human failure

in the face of a storm of this force and scale, is that he lacked imagination. A common failing in a person trained and drilled all his
adult life in military discipline. A common failing in an age hell-
bent on material progress and territorial expansion.

A common failing in any age, perhaps.

It was nearly midnight on Wednesday, January 11, before Lieutenant Woodruff reached a decision about the indications, the term
then in use for the weather forecast, and was satisfied to make it
final. He knew that once he handed the slip of tissue paper to his
assistant, Sergeant Alexander McAdie, there was no going back.
Lives and fortunes depended on his choices, but as a soldier who
had fought in the quicksilver skirmishes of the frontier, Woodruff
was well accustomed to that. He picked up his pen, filled it with
black ink, and scrawled out the forecast for the following day in his
nervous, slanting, but perfectly legible cursive:

> January 12, 12:15 AM—Signal Office War Department,
> Saint Paul.
>   Indications for 24 hours commencing at 7 AM today. For
> Saint Paul, Minneapolis and vicinity: Warmer weather with
> snow, fresh southerly winds becoming variable. For Min-
> nesota: Warmer with snow fresh to high southerly winds be-
> coming variable. For Dakota: Snow, warmer, followed in the
> western portion by colder weather, fresh to high winds gener-
> ally becoming northerly. The snow will drift heavily in Min-
> nesota and Dakota during the day and tonight; the winds will
> generally shift to high colder northerly during the afternoon
> and night.

Woodruff had decided not to issue a cold wave warning. Instructions from Acting Chief Signal Officer Brigadier General Adolphus

W. Greely were extremely clear in this regard. "The exact meaning of the term 'cold wave,'" Greely had written, "implies that the temperature will fall below forty-five (45) degrees, and that in twenty-four hours an abnormal fall of fifteen, or more, degrees will occur." Woodruff himself was something of an expert on cold waves, having written a pamphlet on the subject back in 1885, shortly after he had been detailed for Signal Corps duty. As he well knew, the overwhelming majority of cold waves that hit the Upper Midwest originated east of the Rockies and swept east or southeast down from Montana. Temperatures would plunge first in Helena, then Bismarck and Deadwood in the western reaches of Dakota Territory, then Huron and Yankton in southern Dakota and so on until the cold air reached his own forecast office in Saint Paul. But after studying the 10 P.M. (Eastern time) observations telegraphed from Signal Corps stations to the west, Woodruff concluded that a cold wave warning was not warranted for the next day. Caution was called for, not alarm, especially given how tenuous his position in Saint Paul was. Greely himself had sent Woodruff west to open the office in Saint Paul as part of an experiment in decentralizing the government weather service. Though he had only been forecasting from Saint Paul since October, already Woodruff had issued many more cold wave warnings than his counterparts at the Signal Corps headquarters in Washington had issued the previous year. Better not to cry wolf.

At a few minutes before midnight of January 11, 1888, Woodruff handed the slip of tissue paper with the indications for the twelfth to Sergeant McAdie and instructed him to encode the message and then transmit it by telegraph to the Saint Paul Western Union office, from which it would be distributed to the Office of the Chief Signal Officer in Washington; to the Saint Paul District Telegraph Company; to the Associated Press and the major newspapers in Minneapolis and Saint Paul; to the Signal Corps observers in Milwaukee, Bismarck, Rapid City, and Fort Custer; and to Private Brandenburg of the Minnesota State Weather Service, who would see that it was distributed to sixty-seven volunteer observers in Minnesota and the Dakotas. This was the routine routing procedure for the midnight indications.

Entrusting an officer at a branch office with the task of forecasting the weather was, as both Woodruff and Greely well knew, a bold and radical move. Since 1870, when the Army's Signal Corps first took charge of the nation's weather, all forecasting for the United States had been done by a select handful of "indications officers" working at the Signal Office on G Street near the War Department in Washington, D.C. No matter whether it was a nor'easter bearing down on the New England coast or a persistent series of squalls threatening to flood the Mississippi, all forecasts—initially called "probabilities," then altered to "indications" in 1876—were made in the same circuitous way: Observations were telegraphed to Washington headquarters, maps were drawn and predictions made by the small team of civilian and military meteorologists (most of them commissioned line officers with a few months' training in physics, math, and telegraphy), and the forecasts were then telegraphed back to the field stations as well as to newspapers and railroads. But in the autumn of 1887, under pressure from a group of Saint Paul businessmen worried about the economic consequences of yet another severe winter, Greely agreed to break with Signal Corps tradition and open a branch office in Saint Paul. As Greely wrote later, "The great advantages of knowing sixteen to twenty-four hours in advance that the temperature will fall quickly, apply not only to manifold business interests, but affect the comfort of thousands, and at times the health and life of hundreds." By Greely's estimate, an indications officer in Saint Paul would be able to issue cold wave warnings "from two to five hours" earlier than was possible from Washington.

Assigning the post to Thomas Woodruff was an interesting if somewhat risky move on Greely's part, riskier than the general realized at the time. A handsome well-groomed man with close-cropped fair hair parted in the middle, a bristling Teddy Roosevelt mustache, and a fine prominent nose, Woodruff was a military type more common in the late nineteenth century than in the early twenty-first—an officer and a gentleman. The fact that he was also a weather forecaster was less a matter of personal inclination or talent than a quirk of government bureaucracy and circumstance. Since the Signal Corps suffered from a chronic shortage of officers capable of or interested in observing and predicting the weather, General William B. Hazen, Greely's predecessor as chief signal officer, had started tapping officers from other branches of the Army for Signal duty, generally detailing lieutenants from the artillery, infantry, and cavalry. Relieved temporarily of their other military responsibilities, the lieutenants were dispatched to Fort Myer (near the capital on a site adjoining Arlington National Cemetery in Virginia) for a training course in signaling, electricity, telegraphy, and the basics of physics, math, and meteorology. They were issued sabers and taught to ride horses. They learned how to send messages at all hours and in all weather by flag and torch. They were shouted at and addressed as "fish." They took apart telegraph transmitters to see what made the "click" and put them back together. They were supposed to master the craft of tapping out Morse code. Six months later they emerged as Signal officers. Those select few who showed particular aptitude for the vagaries of forecasting became indications officers.

This was the path that Thomas Woodruff followed starting on February 6, 1883, when General Hazen ordered him to leave his regiment at Fort Keogh on the dry plains of eastern Montana and report to Washington, D.C., for Signal duty.

By this point, Woodruff already had over a decade of strenuous military service under his belt. Like his father before him, he had attended West Point, where he was nicknamed Tim (nicknaming entering plebes is an old West Point tradition) and graduated fifteenth in a class of forty. Immediately after graduation, he signed up with the Fifth Infantry and traveled out to Fort Wallace, Kansas, to join his company. Though he had grown up in Buffalo, New York, and Washington, D.C., with all the comforts and privileges of old Yankee families, the young Thomas Woodruff took to the West at once. For most of his twenties he was on frontier duty fighting "hostile Indians" in Kansas, Montana, and Dakota Territory. In 1876 and again in 1877, Woodruff requested permission to leave safe surveying posts with the Corps of Engineers so he could fight under Colonel (later General) Nelson Appleton Miles in his ruthless campaigns against the remnants of the once great tribes of the Plains—the Cheyenne, the Sioux, the Kiowa, and the Comanche. Woodruff was continually in the field during the autumn of 1877 when Colonel Miles pursued Chief Joseph and his band of Nez Perce holdouts for fourteen hundred miles across Montana. He fought in the five-day skirmish that ended in the capture of the heroic chief ("The Red Napoleon," as the press called him) on October 5, 1877, and he was present when Chief Joseph spoke his famous words of surrender to General Miles: "I am tired of talk that comes to nothing. . . . You might as well expect the rivers to run backwards as that any man who was born a free man should be contented when penned up and denied liberty to go where he pleases."

Despite his bravery in the field, Woodruff harbored no illusions about the glory of combat. He was well aware of the toll these bloody campaigns took on the U.S. military. "These wars are not welcome to the Army," he wrote later. "An Indian campaign means to officer and soldier, toil and hardship, hunger and thirst, heat and cold, imminent danger, perhaps sudden death; or if a man fall wounded, certain torture from which death is a happy release." During the Snow Winter of 1880–81, Woodruff was with Major Guido Ilges fighting the Sioux in Montana. He battled Sitting Bull and his warriors near the Missouri River in the fierce engagement of January 2, 1881. Three Indian villages were destroyed and some 324 prisoners taken (Sitting Bull not among them). It was during that legendary winter that Woodruff recorded a temperature of 63 below zero on his spirit thermometer.

The Indian wars and the extreme weather of the West turned Woodruff into a hardened professional soldier—but the rigors of frontier duty also, rather incongruously, brought out his artistic side. This veteran of some of the most brutal campaigns of the Plains prided himself on his wide reading in literature, history, biography, metaphysics, ethics, and law, and on his accomplishments as a passionate amateur of the gentle arts. "I am very fond of the arts of painting, architecture, and sculpture," Woodruff wrote Greely before he left for Saint Paul. "I have made many sketches from nature in water colors; and also made topographical field sketches, and maps." During his "spare moments" in Indian fighting and surveying, Woodruff made "a large and quite a complete collection of the flora of the 'Staked Plains' [dry mesa country in northwestern Texas and eastern New Mexico], and also a large entomological collection." He published a series of articles describing the Yellowstone Valley and the Bad Lands of Montana in the *Boston Traveller* and an essay on "Our Indian Question" in the *Journal of the Military Service Institution* in which he argued that the United States government's Indian policy was "inconsistent with itself, false in theory, ruinous and cruel in practice, and has in its continued use the ultimate extinction of the Indian race." "Under the banners of civilization and Christianity there have been committed wrongs against the Indian that must cause the most hardened man to blush with shame," wrote Woodruff in 1881, sounding a note that would be repeated again and again over the next century.

Whenever he was granted leaves from military service, Woodruff traveled to Europe. And yet he invariably cut those leaves short in order to rejoin his regiment in its relentless pursuit of "hostile Indians" across Montana and the Dakotas. Though he was convinced that these Indian wars were unjust and unwise, Woodruff did his duty without question. The words *honor, honor-*

able, honor-loving, courage, coolness, duty, value, trust, responsibility, justice, character, honest, and *proper* recur in his writings and correspondence. "Having been educated for the Military profession my desire is to excel in every thing pertaining thereto," he wrote Greely. Woodruff was an officer who evinced and demanded the utmost respect for military discipline, yet he was not embarrassed to set up his easel at the edge of camp as the evening light turned golden or to wander through the Texas scrub, basket and secateurs in hand, gathering samples of switchgrass *(Panicum virgatum)* and blue grama grass *(Bouteloua gracilis)* and the starry yellow flowers of the *Zinnia grandiflora* that bloomed miraculously out of the thin, arid soil. Soldierly and sensitive. Gentlemanly and daring. And *quiet*—that's another word that comrades used to describe Lieutenant Woodruff.

On April 19, 1882, when he was thirty-three years old, Woodruff married Annie Sampson of Cincinnati and obtained a six-month leave from military duties. The newlyweds immediately embarked on an extended tour through England and the Continent, after which Woodruff rejoined his regiment in Fort Keogh, Montana, his wife presumably returning to Cincinnati to live with her parents. The following May, about a month after their first wedding anniversary, Annie Woodruff gave birth to a daughter they named Elizabeth. By then, fortunately, domestic arrangements were considerably easier for the couple since Woodruff had been summoned to Washington in February to report for Signal duty.

When his only child was born on May 9, 1883, Woodruff was enrolled in the officers' training course at Fort Myer, practically within sight of his boyhood home. The following summer he qualified as an indications officer and for the next three years he and his family lived peacefully in Washington, D.C., the lieutenant reporting to work in the Indications Room at the Office of the Chief Signal Officer on G Street while Mrs. Woodruff saw to the duties of running a household and raising a daughter.

"Meteorology has ever been an apple of contention," observed Joseph Henry, the Smithsonian Institution's first director, "as if the violent commotions of the atmosphere induced a sympathetic effect on the minds of those who have attempted to study them." Contention reached a pitch of violence and nastiness inside the Signal Corps during the period of Woodruff's service in the 1880s. Vicious gossip and interoffice backstabbing were rife; charges of incompetence and fiscal impropriety rained down from Congress while Secretary of War Robert Todd Lincoln (the president's son) loudly demanded investigations into the raging scandals; Signal officers routinely aired their grievances in the press; military old-timers railed that the chief civilian forecaster, Cleveland Abbe, known affectionately as "Old Probs," had been brought in over the heads of Army indications officers; civilian employees fumed over the endless snarls of military red tape. All this was bad enough. But there was worse. The chief financial manager of the Signal Corps, a dashing, philandering English-born captain named Henry W. Howgate, was arrested in 1881 for embezzling nearly a quarter of a million dollars (half of which he supposedly spent on prostitutes). Howgate, having absconded when he was released from prison for a single day to visit his daughter, was still at large when Woodruff reported for Signal duty in 1883.

Somehow in this seething atmosphere Woodruff managed to escape all but the mildest slaps and the faintest whispers of error or misconduct. Now and then Woodruff failed to fill out one of the innumerable military forms properly or promptly enough, and he got into a bit of hot water after reporting some malicious gossip that he had heard while inspecting the Chicago Signal Corps station. On one occasion the chief signal officer thundered that his "serious error of judgment" in transcribing the barometric pressure of Salt Lake City spoiled not only "the original chart . . . but an entire day's work of the printed charts." Compared to Greely's routine blasts threatening to sack, transfer, discipline, and/or arrest errant or insubordinate Signal officers, this was very mild indeed. But Woodruff's turn would come.

The fact that violent commotions continued unabated in the Signal Corps after Greely assumed control in 1887 upon General Hazen's death was a grave disappointment inside Grover Cleveland's War Department, for Greely had been appointed expressly to restore order. Indeed, the forty-three-year-old Brigadier General Adolphus Washington Greely looked like the answer to the prayers of those who had despaired of the Corps during the riotous Hazen years. A lifetime soldier, Greely had enlisted as a sixteen-year-old private in the Union Army at the start of the Civil War and moved up steadily through the ranks of the Signal Corps over two decades. He was an ambitious, hard-driving, punctilious officer given to fits of peevish displeasure. His superiors admired and rewarded him for feats like stringing eleven hundred miles of new telegraph lines through harsh, treeless south Texas terrain infested with bandits and hostile Indians and for aggressively recruiting observers for the growing network of Signal Corps weather stations. But it was tragedy and terrible failure that made Greely famous.

On July 17, 1884, First Lieutenant Greely became a national hero when he returned from Greenland—barely alive—with the five survivors of his doomed Arctic expedition. With the financial support and encouragement of the nefarious Howgate—one of Greely's bosom friends—Greely and twenty-four men had set out for the Arctic in the summer of 1881, with the overt scientific mission of conducting research and setting up meteorological stations as part of the First International Polar Year (declared for 1882–83). But like all Arctic explorers before and after, Greely really had his heart set on attaining the pole itself or at least planting the American flag on the "farthest north." This was duly attained when, on May 15, 1882, his second lieutenant, James B. Lockwood, set the Stars and Stripes and a self-recording spirit thermometer in a nine-foot rock pile in western Greenland at a latitude of 83° 24'N, four miles closer to the North Pole than any white explorer had ventured before.

Almost everything else about the Lady Franklin Bay Expedition was a disaster. In 1882 and again in 1883, supply ships failed to reach Greely's camp at Fort Conger on the east coast of Ellesmere Island (northwest of Greenland)—one ship sank with all the provisions; the other, encountering thick ice, returned home with its full cargo of relief supplies. The expedition's orders, drawn up in the comfort of Washington, D.C., specified that in the event that the relief ships never showed up, Greely was to move the party south by September 1, 1883—and on August 9 that's exactly what he insisted on doing, despite the fact that nearly every other member of the crew vehemently objected to leaving the comparatively safe and well-supplied camp at Fort Conger. In the best of circumstances, Greely, when crossed, could be a waspish martinet: He was a dogmatic, stubborn, uncompromising commander who led not by natural authority or earned devotion, but by rigid enforcement of rules and orders. But the rigors of the Arctic brought out his worst. By the time he gave the order to break camp at Fort Conger, most of his men hated him to the point of violence. But they had no choice: Since Greely controlled the supplies, it was either obey or die. After nearly two nightmarish months on drifting pack ice, with winter fast closing in, the party made camp on the desolate wastes of Cape Sabine, some two hundred miles to the south. *"Madness,"* one of the men scrawled in his diary. They had lost several of their boats and much food on the trek south—and there was no resupply cache and no big game to hunt. As the dark frigid months dragged on, they ate their belts and boots and trousers—and then they ate the clothing of the men who died. They ate the filthy oil-tanned covers of their sleeping bags, warming them in a nauseating stew of lichen and seal skin. Finally, in desperation, some of the survivors were reduced to dragging the corpses of the dead out of their shallow ice graves and carving off strips of flesh to swallow in secrecy. In the course of that grim winter and spring, the party's third winter inside the Arctic, eighteen men died—of starvation, of exposure, of suicide, and in one case of a military execution that Greely had ordered as a punishment for stealing food and insubordination. As they starved and froze and watched their comrades die, the men cursed Greely, privately to each other and in their diaries. "This man (I cannot call him a gentleman) comes among us like a serpent in Eden and creates eternal hatred toward himself," one member of the expedition hissed in his diary. "To die is easy, very easy," one of Greely's men scrawled in his diary as life ebbed away; "it is only hard to strive, to endure, to live."

By the time U.S. Navy Commander Winfield Scott Schley reached Cape Sabine on June 22, 1884, only Greely and six of the crew were still alive. "He was unable to stand alone and was almost helpless," wrote Schley of Greely's condition; "all pain of hunger had ceased; his appearance was wild, his hair long and matted, his face and hands covered with sooty, thick dirt; his form wasted almost to a skeleton; his feet and hands were swollen, his eyes were sunken and his body barely covered with dirty and almost worn out garments which had not been changed for six or eight months." One of the six surviving crew members—Corporal Joseph Elison— died a few days later on board Schley's rescue ship after both his legs were amputated. Elison had already lost most of his fingers and both feet to frostbite. He weighed seventy-eight pounds.

Secretary of the Navy William Chandler and General Hazen were on hand to welcome the six survivors of the Lady Franklin Bay Expedition when Schley's ship the *Thetis* steamed into Portsmouth Harbor August 2, 1884. Despite the sensational rumors of cannibalism that dominated the press coverage of the rescue, Greely was accorded the full hero treatment—the military promotion (to captain soon after his return, despite the bitter opposition of Robert Todd Lincoln, who as secretary of war had done everything in his power to thwart the expedition and the rescue efforts); the book deal (*Three Years of Arctic Service* was published by Charles Scribner's Sons in 1886); and, upon Hazen's death, the appointment by President Grover Cleveland to the post of acting chief signal officer, which entailed an automatic promotion to brigadier-general. Greely prided himself on being the first member of the military to rise from volunteer private to the rank of general.

As chief signal officer, Greely moved swiftly to try to clean up the messes left behind by Hazen. After perusing the reports on field stations filed by Signal Corps inspectors (among them Woodruff, who made an extended inspection tour of New England and the Midwest in the summer of 1887), Greely learned just how bad things had gotten. One New England observer was taking nude photos of young women in the weather station. An observer in the Rockies routinely fabricated a week's worth of observations ahead of time and took them to the local telegraph office with instructions to send them off one day at a time so he could spend the week fishing. Yet another observer was forced to hock all the weather instruments in order to pay his poker debts: He made his observations at the appropriate times at the local pawn shop. The observers in New Orleans were extorting regular payments from the local cotton exchanges in return for weather data. Woodruff reported that a sergeant in the Chicago office was convinced that a private had been sent out from Washington to "keep a spy on him." Another observer was, in Greely's words, "foisting useless instruments of his own invention upon this service at an extravagant price." In his first year as chief signal officer, Greely fired a hundred Signal Corps employees.

Greely authorized the experimental indications office in Saint Paul during this flush of reforming zeal, though his choice of Saint Paul as the location had less to do with reform than with politics and pressure from local interests. Indeed, Greely had been lobbied hard that summer by the five prominent businessmen who constituted the "meteorological committee" of the Saint Paul Chamber of Commerce. On August 13, 1887, these gentlemen sent Greely a letter strongly recommending that an indications office be opened in their city (and *not* in Chicago, as Greely himself desired) in order to enhance the timeliness and accuracy of cold wave and heavy snow warnings. About two weeks later, a second and even more urgent request arrived on Greely's desk under the firm, clear signature of one Professor William Wallace Payne, and Payne then followed up his letter with a personal visit to Washington, D.C. This clinched the deal. It's unclear what Greely knew about the character, accomplishments, and aspirations of William Wallace Payne before the business about the indications office—but by the end of that winter he was to know much more than he wanted to about all three.

Payne, a man of "genius and enthusiasm" in the words of one contemporary, was a formidable figure in Minnesota intellectual circles with his finger in many scientific pies. Weather was one of them. Timekeeping another. Hired in 1871 as professor of mathematics and natural philosophy at the fledgling Carleton College in Northfield, south of Saint Paul, Payne persuaded the Carleton trustees to build the first astronomical observatory in Minnesota. Payne then acquired at his own expense a three-inch Fauth transit circle by which he could measure the positions and motions of the stars and planets and two state-of-the-art Howard & Company clocks. Thus equipped he could determine the time more precisely than anyone in Minnesota—indeed, more precisely than just about anyone anywhere in America. This was a matter of no small importance in a country that was rapidly reinventing itself as an industrial and financial giant.

Payne himself strung the wire that connected the readings of his Howard clocks to the nation's telegraphic network and thus put his observatory—and Carleton College—on the map as an official time service. For years the local railroads set their clocks by the signal sent out by Professor Payne, and starting in September 1881,

Payne also relayed the signal for the daily time ball drop in downtown Saint Paul by which residents set their watches (the New Year's Eve ball drop over New York City's Times Square is a vestige of this practice). A fierce rivalry developed between Payne and the U.S. Naval Observatory, which also provided a daily time signal. Not only did Payne pride himself that *his* time was more accurate than Navy time, but he railed against the Naval Observatory for contracting with Western Union Telegraph Company to transmit its time signal around the country. Payne considered Western Union to be greedy, monopolistic, and inefficient, and he never lost an opportunity to blacken the company's name and thwart their increasing control of telegraphy. During the winter of the blizzard, Payne's private campaign against Western Union would figure in his bitter conflict with Lieutenant Woodruff.

With an observatory at his disposal, it was easy for Payne to add meteorological observations to his other scientific endeavors, and in November 1881 he started taking official thrice-daily readings for the Signal Corps (the chief signal officer decreed that these observations be synchronized to Eastern time, with the first observation made at 7 A.M., the second at 3 P.M., and the third at 10 P.M., which meant that observers on the West Coast never got a decent night's sleep). Two years later, Payne became director of the Minnesota State Weather Service—the newly organized Minnesota branch of a largely voluntary state-level network set up for the purpose of gathering weather data, disseminating warnings, and reporting to farmers on conditions affecting their crops. It was in this capacity that he wrote to Greely on August 25, 1887, about the need for an indications officer in Saint Paul.

Greely endorsed Payne's proposal and, for a bureaucrat, acted on it with amazing speed. Within a matter of weeks he had selected Woodruff for the post and he had a member of his staff fire off a letter ordering the lieutenant to "proceed to Saint Paul, Minnesota, and establish in that city, not later than October 20, 1887, an office for the purpose of making weather indications for north western states." The orders stated explicitly that this new office was being established "for the purpose of better serving the stock-growing and other interests in the northwest and with a view of furnishing the information to the public more promptly." Stock-growers first, the public second.

Woodruff was on the road inspecting Signal Corps offices when word of his transfer arrived, and he hurried back to Washington at the end of August to consult with Greely about his new responsibilities. Before departing, he took a ten-day leave to vacation at Nonquitt Beach near New Bedford, Massachusetts, presumably with his wife and four-year-old daughter. It's unclear whether Woodruff brought his family with him to what he later called the "outpost" of Saint Paul, but it seems unlikely. As the sole forecaster for a region larger than New York and New England combined, he would be working for the next six months from nine in the morning to midnight six days a week (with a five-hour dinner break between the afternoon and nighttime observations) with no leave—not a schedule conducive to family life. Even Greely expressed concern that Woodruff would find this relentless routine to be "confining." "It is desired," Greely wrote in his characteristic stiff formal style, "that you shall reduce it [your work] to such an extent that its continuance during the winter will not be detrimental to your general health." Woodruff departed for Saint Paul on October 13 and arrived two days later.

The beginning was rough. Woodruff complained that the Signal Corps office on the top floor of the six-story Chamber of Commerce building in downtown Saint Paul was too cramped and lacked "telegraphic facilities." The furniture was inadequate, the two staff members already on duty—longtime observer Sergeant Patrick Lyons and Private Edwin Brandenburg—were too busy with other tasks to be of much help, and worst of all, the data from other stations that Woodruff needed for making forecasts arrived chroni-

cally late or not at all. A week passed and Woodruff telegraphed
Greely that he was still unable to begin issuing indications. He
scrambled to rent an office in room 60, across the hall from room
55, where Lyons and Brandenburg worked, and managed to install
desks and shelves. Finally, on the night of October 28, Woodruff issued his first set of indications from the Saint Paul office.

The three pages of orders that Greely dictated detailing
Woodruff's new responsibilities included clear and detailed instructions regarding his dealings with Professor Payne: Immediately
after his arrival in Saint Paul, Woodruff was to set up a meeting so
that Payne could brief him on "the general outline of the meteorological work performed by him" as director of the Minnesota State
Weather Service and "in co-operation with the Signal Service." The
meeting went well. Payne wrote to Greely on October 22 that he
was "pleased" with Woodruff and promised to help him. But relations between the two soured very quickly. By November, they had
all but declared war. The conflict flared around the usual issues:
power, authority, money, jealousy, rivalry, control of information. To
begin with, Payne, after his initial welcome, made it very plain that
as director of the Minnesota State Weather Service he considered
himself to have control over the Saint Paul Signal Office, which
meant that Woodruff was his subordinate. "Professor Payne had
quite an idea that I was to report to him," Woodruff wrote later,
"and was quite surprised when I read him the part of my instructions, showing that there was no relation whatever except as to
consulting about the establishment of stations."

Even more rancorous was their fight over how weather data was
to be gathered and communicated, for this directly involved
Payne's old bête noire, Western Union. The two men met to discuss the matter at Woodruff's office on November 14 and, as
Woodruff reported to Greely, Payne "was anything but pleasant."
Back in August, Payne had written Greely that Western Union
might be "willing to allow" the railroads to send weather messages

free of charge, and on the basis of this possibility he urged the general to authorize the opening of twenty new weather stations to improve observations and distribution of warnings in the region. Payne's idea was to expand the meteorological network along the rail lines and get Western Union to foot the communications bill—in essence enlisting the two most powerful and advanced technologies of the day in the cause of better forecasting. That was in August. Now in November, Payne insisted airily to Woodruff that it *didn't matter* what Western Union charged or who had to pay—the messages *must be sent.* Woodruff countered that "as soon as the railroads sent me telegrams of the weather to be used by me in making the general predictions just so soon would the Western Union Company claim the government rates, and that if I received such messages they would be the basis of a claim against the United States." Whereupon Payne demanded, "Is it any of your business or concern where or how messages came provided they gave you information that you could use in making indications?" "It certainly was my business," Woodruff fired back. "I should receive no messages without knowing how they came, and none whatsoever that would in any way compromise the Government and form a basis for a claim for money."

From there the men proceeded to bickering about their relations with the railroads, Payne complaining that Woodruff had gone behind his back in contacting railroad managers about distributing weather reports, and Woodruff retorting that he was under no obligation to include Payne in these negotiations. "I find thus far," Woodruff wrote Greely a week later, "that the Saint Paul and Omaha, and all the other roads do not want to have any or at least wish to have as little to do with Prof. Payne as possible. Whereas in every instance I have been cordially received. . . . The indications and cold-wave warnings are highly appreciated, and are concidered [*sic*] absolutely necessary for the use of Railroad managers, and their subordinates." Woodruff added that the State Weather Service would certainly have received funding from the Minnesota legislature the previous winter "had not Prof. Payne been connected with it."

Payne proved to be a formidable enemy. But this was just the beginning. Woodruff soon found himself fending off a flank attack from the chair of the Meteorological Committee of the Saint Paul Chamber of Commerce, one Thomas Cochran Jr., a local businessman notorious for his shady practices and high-stakes lawsuits. "I find his [Cochran's] standing as a fair and square business man is not good," wrote Woodruff, "though he is leader of several religious organizations." "Not good" was putting it mildly. One of Cochran's former business partners had just accused him of a ten-thousand-dollar swindle (it later turned out that the amount was actually thirty-seven thousand dollars and Cochran was forced to repay it, but by then other lawsuits had been brought against him). It wasn't just Cochran—the entire Meteorological Committee was boiling with scandal and fiscal impropriety. Woodruff learned of an old rumor that the Chamber of Commerce paid out $608 to the State Weather Service in 1886, though no accounting of this sum could be found. And further, Cochran and his cohorts had been circulating a penny postcard to local businesses soliciting "a contribution of $5 to the annual expenses of the Minnesota State Weather Service." The card bore the signature of Private Brandenburg of the U.S. Signal Corps, though Brandenburg strenuously denied having anything to do with this fund-raising scheme.

In December, Greely dispatched a bluff Irish lieutenant on his staff named John C. Walshe (famous in the Corps for pummeling telegraph keys with his fist and pounding rulers to smithereens on his desk) to inspect the Saint Paul Signal office. The general promptly got another earful of dirt. Alluding to the "begging circulars" sent out by Cochran's committee, Walshe reported that "in some way the impression has been produced that the public service here, rendered by the United States Signal Service, depends on the result of money raising by the people, and is a mercenary affair. This is to be deplored, as it distracts from the value of the Action of the Chief Signal Officer in establishing an Indications office as this point." Walshe further noted that Western Union officials complained bitterly that Professor Payne "is continually opposing that company," and that Saint Paul's "very honest, painstaking and conscientious" observer Sergeant Patrick Lyons, "has complained to me that some time ago, Prof. Payne interfered very much with the working of his office."

This was not at all what the sensitive, soldierly Lieutenant Woodruff had in mind when he boarded the train for Saint Paul back in October. He thought he had been summoned westward to provide an important and much-needed public service to the stock growers and ordinary citizens of the Upper Midwest, but instead he found that he had stumbled into a hornet's nest of vicious politics, inflated egos, long-standing feuds, petty turf wars, and unscrupulous business practices. Not that Woodruff himself was blameless in his battle with Payne and the Meteorological Committee of the Saint Paul Chamber of Commerce. A career soldier first and a scientist second, Woodruff made a fetish out of executing his orders punctiliously—even if it meant impeding or sacrificing the pursuit of knowledge. Like all meteorologists, Woodruff was well aware that no matter how accurate his "indications" were, they would be useless if he could not communicate them to the public in time. He knew—again like all forecasters before and since—that the more observed data he had to work with, the better his predictions would be. Yet not only did Woodruff fail to attempt to make his indications available more swiftly and more widely and to expand the network of data-gathering stations, he did everything in his power to block these avenues. The explanation was simple: He had orders from headquarters and he must follow them.

Did Woodruff thereby contribute to the tragedy of the January 12, 1888, blizzard? It's impossible to know for sure. Had Payne gotten his way and set up the twenty new stations, had Woodruff pushed to use even the existing network to get cold wave and heavy snow warnings out faster, had he agreed with Payne that it didn't matter where or how weather "messages" were communicated so long as they *were* communicated, lives might very well have been spared. The science of meteorology, though barely out of its infancy, was advanced enough to predict the intense cold wave that came in the wake of the storm. The technology of the day, though primitive, was sufficient to communicate that prediction all but instantaneously to wherever telegraph wires reached. The fact remains that no one in a position of authority had the imagination or the will to combine science and technology and take action.

By all available evidence, Thomas Woodruff was a fine person, scrupulous about his work, a brave soldier, and a cultivated man of the world, well traveled and well read, liked by his peers and respected by his superiors. Yet it never occurred to him that he might have done something more urgent to alert the people of the Upper Midwest that the mild calm weather of the morning of January 12 would not last. In nothing that he wrote in his official capacity as Signal Corps indications officer does he make a connection between his work and the fate of the people who froze to death or suffocated in blowing snow or lost limbs to frostbite on January 12. The letters and reports he sent back to Washington are full of his resentment of Payne, his meetings with railroad officials, his outrage at local corruption, his bewilderment at the scant public appreciation for "the value of this office and the advantages that it affords." On several occasions he attached to these official reports clippings of articles from the local papers about the work of the indications office and his duties as a forecaster. Yet never once did he allude to the blizzard and its aftermath that occupied the front pages of every newspaper in the region for days.

Which is perhaps another way of saying that the quiet, gentlemanly Lieutenant Woodruff was very much a man of his time and place and rank. Certainly that was how he was judged by his superiors.

Saint Paul in 1888 was a city of red brick and gray granite, steel tracks and cobblestone streets, a solid, young, substantial state capital straddling the northern reaches of the great Mississippi River. To the west, seemingly just over the lip of the bluffs, the endless prairie rolled out, innocent of a single settlement worthy of the name city, with the possible exception of brash, too-close Minneapolis. Fleets of ships crawled up the Mississippi from the older river cities to the south. Some eight hundred freight trains rumbled in and out of the mansard-roofed Union Depot each day along the eighteen rail lines that connected Saint Paul to the bursting industrial and financial hubs back east and the prairie towns that rose along the tracks like columns of smoke. Though it had already developed an incurable habit of glancing nervously at its twin up the river, Saint Paul in 1888 was still booming and proud of it. Some 175,000 citizens (an increase of 200 percent since 1880), ninety-six churches, sixty schools—you'd have to travel clear to San Francisco to find another Western city that could compare.

At the very pinnacle of Saint Paul's boom stood railroad baron James J. Hill. In 1888 the fifty-year-old Hill was roaring ahead with the program of visionary expansion that would earn him the nickname "Empire Builder." He had already transformed his Saint Paul, Minneapolis and Manitoba Railway Company into one of the fastest-growing lines in the country, and the following year he would fold it into the continent-spanning Great Northern Railway Company. It was Hill, more than any other individual, who populated the northern tier states by aggressively luring tens of thousands of immigrants to settle along his rail lines. It was Hill and his Great Northern that put the Pacific Northwest on the commercial map and transformed the dank timber outpost of Seattle into a center of international trade.

As Hill's fortunes rose (he was worth an estimated $63 million at his death in 1916), so rose the fortunes of Saint Paul. The new money that Hill brought in changed the face of the city in the course of the 1880s. Downtown the grid of streets filled in with sprawling block-long commercial buildings and six-story office "towers" tricked out like European palaces and cathedrals, while up on the bluff of Summit Avenue, overlooking the city, the newly rich planted their carpenter Gothic and gloomy Romanesque mansions, some forty-six new houses built between 1882 and 1886. (Hill's own thirty-two-room million-dollar red sandstone pile, completed three years after the blizzard, in 1891, would dwarf every other house on Summit Avenue—indeed every dwelling in the state.)

The downtown Chamber of Commerce building went up on the southwest corner of Sixth and Robert in 1886 during this flush of new construction—a turreted, ornately arched and pilastered neo-Gothic affair vaguely reminiscent of London's houses of Parliament, though on a much smaller scale. At six stories it was among the tallest buildings in the city, making its top floor an ideal location for the Signal Corps observing station. Not only did the sixth-floor windows afford a panoramic view of the sky, but from the top-floor office the observer had easy access to the instruments deployed on the roof. Signal Corps regulations were exact in this regard: With the exception of the two barometers that were to be cloistered indoors and shut up inside their wooden cases when not in use, all other instruments—the four thermometers, the rain gauge, the wind vane, and the anemometer with its four little cups for catching the wind and measuring its speed—were to be set up outside, exposed to the elements (the thermometers had be protected from the sun's rays inside an openwork box). In a bustling city like Saint Paul, a rooftop location was preferred in order to keep the instruments safe from tampering and theft and also because it was thought (mistakenly, as it turns out) that elevating the instruments would produce the best readings.

Residents of the capital were proud to have a weather station atop their new Chamber of Commerce building (and prouder still that the station was *not* in Minneapolis). If they squinted, passersby on Robert Street could just make out the horizontal windmill of the anemometer, whirling in even the faintest breeze. And three times a day anyone who glanced skyward would be treated to the appearance of doughty Signal Corps observer Sergeant Patrick Lyons popping out onto the roof like a cuckoo clock. Quite the natty fellow he was, with clipped dark hair parted in the middle, a stand-up collar and an abundant waxed mustache. Every day of the year in every kind of weather, Lyons or his assistant ventured forth unfailingly to check the rooftop instruments at 6 A.M., 2 P.M., and 9 P.M. local time. Saint Paulites could set their pocket watches by it.

Sergeant Lyons boarded at the Saint James Hotel near the river on Third and Cedar, five long blocks from the Chamber of Commerce building. Five interminable blocks in the dark and bitter cold of winter mornings. But Sergeant Lyons was not in the habit of complaining. A confirmed bachelor in his mid-forties, Lyons had been taking weather observations in Saint Paul for going on fifteen years—just about from the inception of the Signal Corps' weather operations—and in 1888 he had nearly two decades of observations ahead of him. If anyone was used to the weather in Saint Paul, it was Sergeant Lyons.

On the morning of January 11, he reported to work as usual to take the 6 A.M. observations. A light dry snow had started during the night and was still swirling down. The temperature stood at 26 degrees below zero. Lyons noted that it was the fourth morning in a row of double-digit readings below zero.

Lyons was well settled into his daily routine—that day, among his other duties, he climbed out on the roof to oil the anemometer and wipe off the station's battery—when Woodruff and his assistant, Sergeant Alexander McAdie, arrived at the Indications office across the hall at 9 A.M. The men got to work at once. They had an hour and a half before the morning indications were due to be telegraphed to Western Union, and then there were the five daily maps to prepare, copy on the cyclostyle (an early duplicating process using stencil and ink), and distribute to the main local railroads, newspapers, and hotels. Together Woodruff and McAdie went through the data telegraphed from the stations on the regular Signal Corps circuit as well as the additional readings from stations to the north and west that Woodruff had requested. He noted that the atmospheric pressure had dropped sharply at Fort Assinniboine in northern Montana—from 27.31 inches of mercury at 10 P.M. the previous night to 27.06 at that day's 7 A.M. observation. Wind out of the east and barely any temperature change—from -4 to -3. Helena and Fort Custer were both reporting a slight rise in temperature and light winds. Cold and clear in the Dakotas—20 below in Bismarck and Huron—though still not as cold as it was in Saint Paul.

Woodruff and McAdie began entering the 7 A.M. data on blank maps of the United States—for every station in their network, some forty-six stations in all, they filled in the temperature, barometric pressure, wind speed and direction, and state of the weather (fair, clear, cloudy, raining, or snowing). On the first map, Woodruff used red pencil to draw in the isobars, the contour lines connecting stations reporting the same air pressure. As if by magic, evanescent mountain ranges of high and low pressure erupted across the country in ripples and tongues and irregular concentric circles of isobars. Woodruff could see clearly that the huge elongated high centered just north of the Dakotas that had materialized on the previous night's map was shifting southeast. A low seemed to be nudging down from the north behind the high, hence the falling barometer at Fort Assinniboine, but Woodruff did not have readings from far enough north in Alberta or Saskatchewan to get a clear picture of what was happening upstream. The high over the Dakotas was nothing out of the ordinary for this time of year, so it was unlikely that anything really violent was pressing behind it. The sharply dropping pressure over Fort Assinniboine might be a freak or an error or a fabrication by a lazy observer. Woodruff had spent enough time in wind-whipped frontier forts to know that the sergeants and privates who staffed these weather stations were none too reliable. And Fort Assinniboine was one of the most remote military outposts in the country—a vast rectangular treeless compound of two-story brick buildings, second in size only to San Francisco's Presidio, thrown up hastily in the harsh grasslands of northern Montana in the wake of Custer's disaster at Little Bighorn.

Woodruff decided to be cautious. Central Montana stations were reporting slightly warmer conditions, and it was a good bet that this mild weather would spread south and east as the high-pressure center continued to move eastward across the country. He wrote out the afternoon forecast by hand in black ink on a slip of tissue paper:

January 11, 1888—10:30 AM
    Indications for 24 hours commencing 3 PM today.
    For Saint Paul, Minneapolis and vicinity: Slightly warmer fair weather, light to fresh variable winds.
    For Minnesota and Dakota: Slightly warmer fair weather, light to fresh variable winds.

Woodruff would keep an eye on those Montana pressures when he issued his next set of indications at midnight. Though the press and the public still believed that forecasting the weather was more hocus-pocus—or hoax—than science, Woodruff was convinced that the essential elements were clear and straightforward. If you could define the areas of high and low pressure, identify their centers, and track their movement, you could pretty much predict the rise and fall of temperature and the likelihood of stormy weather over the next day and often over the next two days. As Woodruff wrote in his paper "Cold Waves and Their Progress," "In various investigations and studies, it has been shown that a fall of temperature succeeds or follows an area of low barometer, and a rise precedes such an area; and that, in general, the reverse is true of an area of high barometer, viz.: that a fall precedes and a rise follows it." In other words, temperatures rise in advance of an approaching low-pressure system and fall once the low has passed, while high pressure causes temperatures to fall as it builds in and rise as it breaks down. Woodruff was aware that highs and lows "move almost invariably across the United States from west to east," and further, that the movement of a low seemed to determine the movement of the high-pressure area following behind it, almost as if low pressure dragged high pressure along in its wake. "[O]ne storm begets its successor," wrote Elias Loomis, a pioneer of American weather science whose *Treatise on Meteorology* Woodruff had read during his training course at Fort Myer. "The undulations thus excited in the atmosphere bear considerable analogy to the waves of the ocean agitated by a tempest, and which are propagated by mechanical laws long after the first exciting cause has ceased to act."

Woodruff concluded from researching his paper that the cold waves originating in "the vast regions of ice and snow near the arctic circle" almost always entered the United States through Montana, and from there took one of three tracks: due east "along the chain of great lakes and across New England"; southeast into Dakota, Nebraska, Iowa, Missouri, and so on across the entire country; or due south through Dakota Territory and all the way down to Texas, where the cold air sometimes veered northeast and spread up the Atlantic coast. The critical question for the forecaster was: What made a given cold wave take a particular track? Woodruff frankly admitted ignorance. "Even after a decided cold wave is observed in the extreme northwest," he wrote in his paper, "we are not able to determine which one of the three paths it will take." Greely conceded much the same thing, though rather more pompously, in his book *American Weather*: "As yet it has not been determined with absolute accuracy what conditions must obtain to induce the passage of cold waves" in one direction over another. "The question doubtless depends upon the relative relation of the centre of the anti-cyclone to that of some cyclone far distant." Since they had no knowledge of fronts and their role in structuring storms and only the vaguest idea of how upper-air conditions influence what happens on the ground, forecasters of the day fell back on probabilities based on statistical analysis of existing data—and guesswork. Statistics told Woodruff that over half the cold waves entering the U.S. from Helena moved southeast, while only about a quarter plunged directly south down to Texas; and further that 71 percent of these Helena cold waves hit Bismarck in eight hours and 88 percent reached Omaha within a day. Interestingly, Woodruff also ascertained that nearly half of the cold waves in his sample were first detected at the 3 P.M. observations (2 P.M. Central time)—typically the warmest time of the day in winter—an insight that Greely mentioned (crediting Woodruff) in *American Weather.*

It wasn't much to go on in the face of the tremendous surge of energy spinning down at him from the north. A couple of formulas, a few charts of statistics, some rather murky mumbo-jumbo about the three possible paths. Even with a rooftop bristling with instruments and the all-important telegraph wires connecting him to the national grid, what Woodruff saw when he looked out from his office at the Saint Paul Chamber of Commerce building was more a mirror of his own mind than a window on reality.

※ ※

Woodruff and young Alexander McAdie broke for lunch as usual at 1 P.M. on January 11 and were back in the office at 2:30 P.M., Central time—half an hour after Lyons took the afternoon observations (the temperature in Saint Paul had risen 10 degrees since the morning readings to a comparatively balmy 15 below zero). One hundred and thirty-two other Signal Corps sergeants and privates were taking (or faking) observations at precisely the same time—3 P.M. in Washington, D.C.—but it would be two hours before the data arrived in Saint Paul from the Chicago Western Union office (to hasten distribution, a circuit was established through Chicago so that data from stations throughout the West could be transmitted simultaneously and then relayed to Saint Paul). And then it took McAdie another hour to translate the telegrams (weather messages were telegraphed in code to make them shorter—and thus cheaper to send—and to avoid numerical errors). Back at the Signal Office in Washington, they had the transcription and mapping process down to a science: as the telegrams came in on the wires from all over the country, a team of clerks read aloud the translations of the cipher while another clerical team recorded the new data, each clerk working on only one stream of data (temperature, air pressure, and so on). Meanwhile, the indications officers hovered behind the clerks and watched the pressure systems emerging on the maps, so that by the time all data were entered and all the isobars drawn, they were ready to issue their forecasts. Very likely McAdie and Woodruff worked in a similar fashion in Saint Paul, though necessarily more slowly since there were only two of them (or three when Private W. H. Ford was on duty).

It was approaching 5 P.M. in Saint Paul before Woodruff got a clear picture of what had happened in the course of the day. The readings from Fort Assinniboine were startling. The pressure had fallen dramatically since the 7 A.M. observations—from 27.06 to 26.76, a drop Woodruff had rarely seen in all his years in the West—while the temperature had risen 11 degrees, from 3 below to 8 above. Bismarck was also reporting a rapidly falling barometer and rising temperatures. Farther south in Huron, a stiff southeast wind had kicked up and the temperature had jumped 18 degrees, from 20 below at 7 A.M. to 2 below at 3 P.M.

As Woodruff inked the isobars in red on the map, a distinct oval bowl of low pressure took shape around Medicine Hat in southern

Alberta, just north of Fort Assinniboine: the first faint shadow of the coming storm.

Woodruff and McAdie left the office at 5:45, shortly after tabulating the afternoon observations. The temperature was still rising in Saint Paul, odd for this time of day in winter, though with readings remaining in the double digits below zero, they hardly noticed. The men took four and half hours off for supper and a rest, and then returned to the Chamber of Commerce building at 10:15 P.M. to await the arrival of the nighttime observations from Chicago. Again there was the time lag due to the backlog at Western Union and then the tedious process of translation and transcription. Woodruff worked quickly on the charts—he and McAdie were eager to get home—and the highs and lows bloomed under his hand like faint red targets.

He saw at once that the Alberta low from the afternoon chart had moved quickly to the southeast and was now centered over Fort Keogh, the eastern Montana army post where he had been stationed with his regiment when he was called up for signal duty. But still no outbreak of truly frigid air over the U.S. Fort Assinniboine reported 4 above at the 10 P.M. (Eastern time) observation with slightly rising barometer and northerly gales. Helena was 33 above, Bismarck 7 above, and the Huron station, staffed since 1881 by the conscientious Sergeant Samuel W. Glenn, was reporting a slightly falling barometer, stiff southeast winds, and a temperature rise of 5 degrees in the past seven hours, from 2 below to 3 above.

The reading that stood out most starkly was the temperature at North Platte, in southwestern Nebraska : 22 degrees above zero at 10 P.M., a rise of 20 degrees since the 3 P.M. observations.

It took Woodruff only a few minutes to complete the midnight indications. It was perfectly clear to him from the 10 P.M. observations that a deep low was in the process of dropping southeast over the country—the classic path. He knew that temperatures would rise in advance of the low and fall once it had passed. He knew that Helena held the key. Once temperature began to drop in Helena, they would be likely to drop in Bismarck eight hours later (71 percent of the time according his calculations) and in Saint Paul within twenty-four hours (73 percent of the time). But as of 10 P.M., Helena was still showing a *rise* of temperature—so there was as yet no cause for alarm. This rise in Helena was a bit puzzling, because according to Woodruff's own formulation "a *fall* of temperature succeeds or follows an area of low barometer," and the center of low pressure had already passed well to the south and east of Helena. Woodruff undoubtedly believed that this inconsistency would have been resolved if he had access to more timely data from a more extensive network of stations. As he had written in "Cold Waves and Their Progress," the difficulty of forecasting "is increased by the fact that observations at our Signal Service stations are separated by intervals of eight hours." Eight hours was an eternity meteorologically, especially in the wild and mercurial West.

But there was one critical bit of data right there in front of him, if only he knew it. That 20-degree spike in temperature at North Platte was like an arsenal packed with explosives. The meaning of this reading was lost on Woodruff, as it would have been lost on Greely or Professor Abbe back in Washington. It was one more piece of data that didn't fit what Greely termed the "principles of philosophy" that he insisted were "sufficient to explain the intricate and varied phenomena of the atmosphere."

By 11:45 P.M. on January 11, the indications for the following day were obvious to Woodruff. The Montana low would bring snow and rising temperatures to Dakota Territory during the day, followed by colder temperatures and northerly winds spreading over the region from west to east. The tight rings of isobars bunched together over southeastern Montana, northeastern Wyoming, and western Dakota indicated that pressure differences were particularly marked in this region—a clear sign of high winds. Woodruff thus predicted that sometime during the day the snow would drift heavily as stiff winds blew from the north in the wake of the low.

No cold wave warning, however, was called for—there was no justification for it with temperatures of 33 above in Helena.

⚜ ⚜

When Woodruff and McAdie left the Saint Paul Signal Office a few minutes before midnight, E. J. Hobbs, the Signal Corps observer in Helena, was still awake and still on duty at the station. Word had reached him of severely cold weather bearing down from the north. Even though he was under no obligation to remain at his post, he had decided to stay up all night to monitor the situation. Strangely, temperatures continued to rise through the first hours of January 12. At 2:30 A.M., Hobbs recorded 38 above and the mercury hit 40.5 before dawn.

Meanwhile 173 miles to the northeast, inside the long, sturdy brick structure that housed the Fort Assinniboine weather station, the Signal Corps observer on duty was observing something he had never seen before. The telegraph wires that connected the station's receiver to the endless loops of wire strung over the plains were emitting strange flashes of light—"a constant play of light," as the private noted in the station's journal—and were so charged with electricity that they could not be handled. Even inside the station the air fairly crackled with electricity. And outside the weather was turning fiercer by the minute. The warm chinook wind that had blown out of the southwest all day was now howling out of the northwest at a velocity approaching 50 miles an hour.

It was like a hurricane approaching over the empty plains of Montana in the dead of winter.

CHAPTER FIVE

Cold Front

Of course, Lieutenant Woodruff and E. J. Hobbs in Helena and Sergeant Glenn in Huron knew what a cold front on the prairie *felt* like—the sudden shift in wind direction, the burst of rain or snow, the crystalline, head-clearing blast of cold air from the northwest. Woodruff even used the term "the front of the wave" in "Cold Waves and Their Progress" to describe how cold waves tend to track the paths of lows from northwest to southeast across the country. But it wasn't until after World War I that a team of Norwegian meteorologists led by the physicist Vilhelm Bjerknes and his son Jacob zeroed in on the importance of cold and warm fronts in the structure of weather systems and devised a way of graphically representing their location and direction on weather maps. The insight of Vilhelm and Jacob Bjerknes nudged the fledgling science of meteorology one large step closer to maturity.

The Bjerkneses borrowed the term "front" from the vocabulary of the war that had just decimated Europe and brought neutral Norway to the brink of mass starvation. In World War I, the fronts were the long, wavering lines where the two opposing armies met, dug in, fought, and advanced or retreated after terrific violence. The analogy, the Bjerkneses realized, fit weather exactly. Air masses at a front come together too rapidly to mix. Instead, at a cold front, an advancing mass of dense cold air shoves itself under warm air, forcing the warm air to rise rapidly along a steep incline and condense out its moisture: The speed and abruptness of the ascent typically results in short-lived bursts of heavy rain or snow, often accompanied by lightning. Warm fronts, in which warm air advances up and over cold air, progress more gradually and tend to bring lighter but steadier and longer-lasting precipitation to a wider area. Fronts are the seams in the atmosphere that extend from the ground to the tops of clouds—long, rippling, fragile seams that get ripped apart by storms.

A colleague of the Bjerkneses in Bergen later came up with the idea of representing fronts on maps as lines of sawteeth—triangular barbs for cold fronts and soft semicircular pips for warm fronts. Today it's hard to imagine the weather map without the warm and cold fronts snaking across it—but this elegant bit of shorthand has only been around since the 1920s. There were no fronts on the maps Lieutenant Woodruff drew and distributed every morning to the eager citizens of Saint Paul.

An evil genius could not have devised a more perfect battleground for clashing weather fronts than the prairies of North America. When conditions are right, which they frequently are, vigorous fronts unleash the worst weather in the world over this region—super-cell thunderstorms spawning tornadoes in late spring, huge globes of hail falling from anvil-topped cumulonimbus clouds in summer, blizzards in winter. On the prairie, cold fronts can come through so rapidly that standing water ices up in ridges, small animals literally freeze to their tracks, people whose clothing is wet find themselves encased in ice. When a strong cold front is accompanied by the blowing icedust of a blizzard, the punishment inflicted on human and beast is unimaginable.

While the slip of tissue paper with Lieutenant Woodruff's indications for January 12, 1888, sat on a countertop at the Saint Paul Western Union at 113 East Fourth Street waiting to be translated into Morse code and sent out along the wires to newspapers and Signal Corps offices, one of these ferocious cold fronts was dropping southeast through Montana at around 45 miles an hour. With the advance of the cold front, all of the elements of the storm suddenly began to feed off each other, bloating up hugely with every bite. As the contrasting air masses slammed together, they caused the upper-level winds to strengthen, which served to strengthen the low. As the low deepened, the surface winds increased, causing the temperature differences between the air masses to spike. The greater the temperature difference, the faster the low deepened. The deeper the low, the stronger the front. It was a self-reinforcing and accelerating cycle.

When a storm becomes organized gradually, high wispy cirrus clouds usually appear a day or two ahead of the cold front, followed by a solid low bank of stratus cloud stealing across the sky. But by the first hours of January 12, this storm was spinning up so quickly that there was no time for an atmospheric herald. The cold front was now so strong and so well defined that it was like a curtain of ice separating two radically different climates, a curtain that was hurtling in two directions simultaneously—down from the sky and horizontally across the surface of the earth. At the same time that the curtain swept down from the north, a warm spongy mass of air was ascending from the opposite direction. The intensifying low forced the two air masses to converge with ever increasing speeds. When they collided, the atmosphere erupted.

The warm air slid up and over the curtain, rising about three feet every second. As soon as it hit an altitude of about 5,000 to 7,000 feet, the air instantly surrendered its vapor into infinitesimal droplets of supercooled water—liquid specks colder than freezing but prevented from turning to ice by the surface tension of their "skin." As many as a billion cloud droplets swarmed around every cubic meter inside the ballooning clouds. Even smaller particles of airborne debris roiled alongside the cloud droplets—pollen, dust, salt crystals, shredded spiderwebs—and these particles in their billions served as the nuclei around which the supercooled cloud droplets coalesced and turned to ice. The instant the cloud droplets froze, they began to grow exponentially by fixing other droplets onto their crystalline facets.

Behind the front, where the air was much colder, ice production inside the clouds happened at significantly lower altitudes, as low as 3,000 feet, and the condensing vapor spat out a different kind of crystal. It was too cold to form the lacy star-patterned crystals known as dendrites—the pretty snowflakes of Christmas cards. Too cold for the little pellets of graupel that accrete as ice crystals glue on layer after layer of supercooled droplets. Instead, what was being manufactured inside these frigid clouds was a myriad of nearly microscopic hexagonal plates and hollow columns and needles—hard slick-surfaced crystals that bounced off each other as they swirled around.

The snow that fell when these plates and columns grew heavy enough was hard as rock and fine as dust. Actually "fell" gives the wrong idea. The plates and needles and columns billowed out of the bases of the clouds in huge streaming horizontal veils, as if the bank of icy clouds had descended to earth and burst apart in the gale. The newly manufactured snow crystals, smashed and ground into infinitesimal fragments within seconds of their creation, mixed with older crystals that had settled at the surface after previous storms. In the gale the crystals of different vintages became indistinguishable.

New snow is not necessary to boost a winter storm into the category of blizzard. All that's required is wind of at least 35 miles an hour, airborne crystals, and temperatures of 20 degrees or colder (the National Weather Service recently dropped the temperature requirement). The January 12 storm qualified easily.

The disturbance rippled southeast on precisely the path Woodruff had expected, the path taken by 54.14 percent of the cold waves he had studied. By midnight, the leading edge of cold air had reached Poplar River in northeastern Montana. By 2 A.M., it had engulfed Medora in western Dakota, where Teddy Roosevelt had lost so many cattle to blizzards the previous winter. By 4 A.M., January 12, the cold front was poised just west of Bismarck.

At six o'clock in the morning Central time, when the telegrams bearing Lieutenant Woodruff's midnight indications began to arrive on the desks of weather observers and newspaper editors, the temperature at North Platte, Nebraska, stood at 28, fully 30 degrees warmer than the previous day; while in Helena, 670 miles to the northwest, observer E. J. Hobbs, who had stayed up all night, noted that the mercury had fallen 49.5 degrees in the past four and a half hours, from 40.5 above to 9 below zero. Omaha was 23 above and so was Yankton in southern Dakota. The report from Huron, northwest of Yankton, was late—but eventually this bit of data got entered on the map: 19 above at 6 A.M. Central time, nearly 40 degrees warmer than the previous day.

More fuel for the approaching storm.

Of course, nobody in Nebraska or Dakota saw it that way. The tragedy of that day was that the fuel came disguised as welcome relief from the weeks of bitter cold, as a spell of softness and relative ease in lives that had too little of either, as an invitation to get outside and work or walk or simply breathe before the weather closed in again.

In the event, word reached the settlements of southern Dakota not from the Signal Office, but from telegrams sent over the railroad lines. In Codington County in eastern Dakota, just two hundred miles west of Saint Paul, a station agent named Brown received a telegram from Bismarck around 8 A.M. reporting "rapidly falling temperatures and an arctic gale which was steadily increasing in violence." There was a schoolhouse near his station, and when Brown saw the children walking past he rushed out at them and stood in the middle of the crossing. He shouted at the kids that "the worst blizzard of the season would be there in two hours." Many children went back home. All of those who continued on to school regretted it.

CHAPTER SIX

Explosion

An hour after station agent Brown received the telegram from Bismarck warning of the storm, Lieutenant Woodruff arrived at his office in downtown Saint Paul. Sergeant Lyons reported smartly that it was the warmest morning of the week—2 below at the 6 A.M. observation, fully 23 degrees warmer than the day before. Woodruff and McAdie got to work immediately on the morning indications. They had an hour and a half to go through the stacks of telegrams from the outlying stations, draw the five maps, prepare the cyclostyle—the same routine as the day before, the same as the day after.

Woodruff was gratified to see that his indications from the previous night had "verified," particularly in the northwest. Temperatures had indeed risen in advance of the low, just as he anticipated. And behind the low, up in Montana, cold air was spilling out of Canada—2 below zero at Fort Assinniboine with winds blowing out of the north at 28 miles an hour. Eight below and snowing in Helena.

There was no question now in Woodruff's mind: a cold wave warning was warranted for Dakota and for Nebraska later in the day. It would blow hard and the snow would drift heavily and then the temperature would fall, in a long arc sweeping from northwest to southeast, the classic path for an advancing cold wave. By nightfall, western Dakota would be scoured by arctic winds. Woodruff knew what it felt like when that blade of cold penetrated. He had not forgotten his winters at Fort Keogh.

Woodruff took a fresh sheet of tissue paper and quickly covered it with black ink:

Signal Office War Department
Saint Paul Minnesota
January 12, 1888, 10:30 AM
   Indications for 24 hours commencing at 3 PM today:
   For Saint Paul, Minneapolis and vicinity: warmer weather with snow; fresh easterly winds.
   For Minnesota: snow warmer followed in northern part by colder fresh to high variable winds becoming northerly
   For Dakota: Snow warmer followed by colder with a cold wave, fresh to high northerly winds
   A cold wave is indicated for Dakota and Nebraska tonight and tomorrow; the snow will drift heavily today and tomorrow in Dakota, Nebraska, Minnesota and Wisconsin.

The words *cold wave* in the indications triggered a set of special procedures. The instructions were clear and exact in the military way. Newspapers and the Associated Press wire service received the daily indications as a matter of course, but now extra telegrams must go out to the Signal Service stations in Minnesota, Dakota Territory, Nebraska, Iowa, Wisconsin, and Chicago, some twenty-two stations in all, as well as to the principal railroads serving the region. Professor William Payne had also arranged with Woodruff to have cold wave telegrams sent to the sixty "flag stations" that he had set up through his Minnesota State Weather Service; although, as Payne noted sourly, "the service as rendered by the Western Union Telegraph Company, in many instances is very poor," with the result that his volunteer flag displaymen frequently did not receive Woodruff's indications or received them late.

In any case, when and if the messages arrived, displaymen at the flag stations and observers at the Signal Corps observation stations were immediately to hoist the black and white cold wave flags— six-by-eight-foot white rectangular sheets with a two-foot black square centered in the middle—and keep them flying until Woodruff instructed them to take them down.

That was the procedure and on January 12 it worked as well as it was expected to. The forecast was substantially correct. The messages were coded and transmitted and duly received. The orders were obeyed. Word went out—the official word as sanctioned by the Signal Office of the War Department.

But by the time the procedure went into effect, it was too late to matter.

In the central Dakota boom town of Huron on the frozen James River, Signal Corps observer Sergeant Samuel W. Glenn was violently ill that morning—so ill that he was late getting the 7 A.M. observations off to Woodruff in Saint Paul and Greely in Washington. This was a most unusual occurrence. Sergeant Glenn was not one of those slipshod, shady observers who pawned the barometer to pay off poker debts or shook down local businessmen for cash in exchange for weather data. Far from it. For Glenn monitoring the weather was both a career and a mission, and rare was the occasion when he failed to perform his duty as the observer in charge of the Huron Signal Corps station punctually and precisely.

To the extent that a town as young and raw as Huron could have

institutions, Sergeant Glenn was one of them. It was he who inaugurated the downtown office on Third Street near the opera house
six and a half years before, taking his first observation at 5:35 A.M.
on July 1, 1881, when the town, in the words of one settler, was
nothing but "long lines of weather beaten square-fronted stores,
tar-paper covered shacks with one way roofs, sod houses, tents,
wagon camps, saloons galore, no churches or schools, and streets
hub deep in mud most of the time." Since the Signal Corps building was then still under construction, Sergeant Glenn had to climb
a ladder to get up to his office. Now, some 118,500 observations
later, Glenn was still at it—five observations a day (the three standard Signal Corps readings, plus two more for local records) taken
with his four thermometers, two barometers, one anemometer with
a self-registering attachment, one anemoscope, rain gauge, tele-
scope, and field glass. Sergeant Glenn kept his arsenal of instruments in perfect order, wrote meticulously and faithfully in the
station's journal of any unusual local meteorological phenomena,
and received the highest praise from the Corps inspectors dispatched each year from Washington. In fact, Lieutenant John C.
Walshe had inspected the Huron station on November 26 and 27,
shortly before meeting with Woodruff at the Indications Office in
Saint Paul, and pronounced it exemplary.

Glenn's lapse on the morning of January 12 actually had less to
do with his illness than the cure. As he noted in the station journal,
he been sick on and off since January 3 with an undisclosed ailment
and had been receiving "medical advice and medicine from Dr. Alford." Before dawn on January 12, Glenn had unwisely taken too
much of Dr. Alford's medicine on an empty stomach—alcohol may
well have been the primary ingredient—and the reaction was swift
and terrible. While Glenn languished in bed, somebody, presumably the station assistant, dragged to the office at 6 A.M. Central
time to check the thermometers and rain gauge outdoors and the
barometers and anemometer register mounted on the station wall.

The observations, though sent late, do exist, and Glenn was far too upstanding an observer to have fabricated them. Atmospheric pressure 28.21 inches of mercury and falling; temperature 19 degrees; wind out of the south at 24 miles an hour. Though Glenn was an hour late getting the telegrams off, not a single reading was missing. (He later supplied headquarters with a certificate signed by Dr. Alford attesting to his illness.)

By midmorning, the effects of Glenn's medicine had worn off sufficiently for him to climb out on the roof of the Huron station. Having noted the rapid fall in the barometric pressure, Glenn concluded that the region was in for a gale. Before it hit he wanted to inspect the anemometer to make sure the wires connecting it to the register inside were in good order. He also, frankly, wanted to see the storm blow in, always a spectacle out on the prairie. And so at 11:42 A.M. Central time Sergeant Glenn went out on the roof to have a look around.

Sixty seconds later, he came within a whisker of getting blown off.

Glenn must have had his pocket watch open in his hand because he recorded to the minute in the station's journal what occurred in the atmosphere in the moments that followed: "The air, for about one (1) minute, was perfectly calm, and voices and noises on the street below appeared as though emanating from great depths. A peculiar 'hush' prevailed over everything. In the next minute the sky was completely overcast by a heavy black cloud, which had in a few minutes previously hung suspended along the western and northwestern horizon, and the wind veered to the west (by the southwest quadrant) with such violence as to render the observer's position very unsafe. The air was immediately filled with snow as fine as sifted flour. The wind veered to the northeast, then backed to the northwest, in a gale which in three minutes attained a velocity of forty (40) miles per hour. In five minutes after the wind changed the outlines of objects fifteen (15) feet away were not discernible."

The Signal Office in Washington supplied observers like Sergeant Glenn with thick volumes of lined blank pages in which to keep the station journals. Months of the year and days of the month were preprinted on the quarto pages, three days per page with about three inches of blank space for each day. Sergeant Glenn wrote "disastrous storm" in the margin next to the three inches allotted to January 12, 1888, filled the blank space to overflowing, and then, at his own initiative, pasted into the journal nine additional handwritten pages. It is an invaluable account of the violence that tore the atmosphere over Huron starting at 11:42 A.M. that day and the suffering that living creatures on the ground endured for weeks afterward.

※ ※

Sergeant Glenn had his two barometers to alert him that something powerful was coming his way. But out on the prairie and in the one-room country schoolhouses and along the flimsy, flammable main streets of the quick-built railroad towns, the blizzard took people utterly by surprise.

To those standing outside, it looked like the northwest corner of the sky was suddenly filling and bulging and ripping open. In account after account there runs the same thread, often the same words: There had never been anything like it. Settlers who had lived through the blizzards of 1873 and the recurring storms of the Snow Winter of 1880-81 and the vicious blizzards that had killed so many cattle just the previous winter—none of them had ever seen a storm come up so quickly or burst so violently.

"My brother and I were out snowballing on a bank," remembered Allie Green, a fifteen-year-old boy in Clark County in eastern Dakota Territory. "We could see the blizzard coming across Spirit Lake. It was just as still as could be. We saw it cut off the trees like it was a white roll coming. It hit with a 60 mile an hour wind. It had snowed the night before about two or three inches. It just sucked up that snow into the air and nearly smothered you."

It was like a "gray wall," said H. G. Purcell, a schoolboy in neighboring Codington County, who stood in awe on a ball field at the edge of town as the storm bore down from the northwest.

"We were all out playing in our shirt sleeves, without hats or mittens," remembered a South Dakota schoolboy. "Suddenly we looked up and saw something coming rolling toward us with great fury from the northwest, and making a loud noise. It looked like a long string of big bales of cotton, each one bound tightly with heavy cords of silver, and then all tied together with great silvery ropes. The broad front of these cotton bales looked to be about twenty-five feet high; above them it was perfectly clear. The phenomenon was so unusual that it scared us children, and several of us ran into the schoolhouse and screamed to the teacher to come out quickly and see what was happening." When the storm reached the schoolhouse a few moments later, it hit "with such force that it nearly moved it off its cobble-stone foundation. And the roar of the wind was indescribable."

"The sky was inky," wrote a teacher at the Rosebud Indian Agency in Dakota, just north of the Nebraska line. "Lilia [another teacher] ventured a few yards out of the front door at its beginning, and was near not getting back. The wind struck her with such violence as to bring her head down to her knees, and take away her breath. She said she was near falling on her face, and she knew that if she fell she would not get up again."

Norris E. Williams, a schoolteacher in Jerauld County, west of Sioux Falls, was standing in front of his schoolhouse with a group of students during the late-morning recess when the storm descended: "I was just saying that I ought to dismiss school and go to Woonsocket for coal when a sudden whiff of cold air caused us all to turn and look toward the north, where we saw what appeared to be a huge cloud rolling over and over along the ground, blotting out the view of the nearby hills and covering everything in that direction as with a blanket. There was scarcely time to exclaim at the unusual appearance when the cloud struck us with awful violence and in an instant the warm and quiet day was changed into a howling pandemonium of ice and snow."

Darkness fell, "darkness that might be felt," as one farmer wrote. "You could hardly see your hand before you or draw your breath and that with the intense cold roaring wind and darkness it would appall the stoutest heart."

꧁ ꧂

Many wrote that the onset of the storm was preceded by a loud roar, like an approaching train. It was a roar they not only heard but felt vibrating in their gut.

That sound was the wind at the knife edge of the cold front smashing the snow to powder. Dr. Louis W. Uccellini, director of the National Centers for Environmental Prediction and an expert on snowstorms, compares what was happening on the ground as the cold front came through to the smoke and ash roiling through the canyons of lower Manhattan after the towers of the World Trade Center collapsed on September 11, 2001: "It was not a laminar flow in which the currents move in parallel layers, but a flow moving in turbulent eddies. The turbulence behind this front must have been incredible. The air was rolling over at the same time that it was coming down. The effect was like putting the snow and ice in a grinder. The turbulence pulverized the snow to talcum powder as it entered the last mile or so of the atmosphere above the ground."

꧁ ꧂

That morning the four classrooms of the Groton School were full for the first time since Christmas on account of the balmy weather. Perhaps a hundred children in all had walked from the houses lining Main Street to the two-story schoolhouse set optimistically beyond the fringe of settlement. As usual, eight-year-old Walter Allen had gotten to school early. As the row monitor of his classroom, he had important duties to perform. And so even before his father W.C. Allen left for his law office next door on Main Street or his two older brothers headed out to their jobs at the sole remaining Groton newspaper, Walter ducked behind the family's house and cut across the backyard toward school.

On a mild morning like January 12, the walk from the Allen house to school was nothing, even for an eight-year-old. As long as Walter stayed away from the deep drifts that massed behind buildings and fences, he was fine—and anyway, once he got beyond the outhouses tucked discreetly in back of the Main Street homes there were few structures of any kind until he reached the school. Not a single tree varied the monotony or broke the wind, though for once that morning the wind wasn't a problem.

The vestibule of the school where the children hung their wraps and stowed their wet shoes was less jammed than usual since the weather was so mild. Some kids had dispensed with cloaks and overcoats altogether—their wool stockings and homemade linsey-woolsey and calico dresses and petticoats or thick shirts and trousers would have been warm enough on such a mild morning. One girl remembered slipping out of the house in a short cape made from the bottom of an outgrown coat before her mother spotted her and insisted on the heavy coat. "My brothers wore little homemade denim jackets. No scarves, mittens or overshoes, for it promised to be a fine winter's day. Long before I reached school, I was carrying my cape in my hand." It was hard on a day like that to be in school at all, but finally the teachers managed to herd all of them inside and get them settled—children ranging in age from five or six to fourteen, divided among four classrooms in the two-story frame schoolhouse.

Walter in his coveted front-row seat opened his desk and took out his slate, his erasing rag, and his prize possession—a delicate little perfume bottle. All the kids kept corked bottles of some sort in their desks to use in cleaning their slates—they'd pour a bit of water from the bottle onto the slate and scrub it clean with a rag. Walter cherished the glass perfume bottle not only because it was so pretty and fine to look at, but because the stopper in its neck allowed him to squirt out just a few drops at a time. No one else in his classroom had anything like it.

The morning began like every other with recitation and chanting in unison of a passage from their "reader"—very likely one of the McGuffey readers that were then almost universal throughout the Midwest.

> Hear the child-ren gay-ly shout,
> "Half past four, and school is out!"
> See them, as they quick-ly go,
> Trip-ping home-ward o'er the snow.
>
> Thus these lit-tle chil-dren go,
> Trip-ping home-ward o'er the snow;
> Laugh-ing, play-ing, on their way
> Ver-y hap-py, glad, and gay.
>
> There are many children whose parents are too poor to send them to school. Do you not pity them? Take good care of your new book, and give your old Reader to some child who is too poor to buy one.

Around 10:30, the chanting in Walter's classroom abruptly ceased. Everyone had gotten up to look out the windows at the sky. The windows and doors rattled as the wall of shattered crystals slammed into the school. In an instant the droning orderliness of the classrooms dissolved. Walter watched his teacher leave the room to confer with the other teachers and then return quickly. They had decided to dismiss school for the day and send the children home. Walter's teacher had to shout over the roar of the wind that the row monitors were to get the wraps and distribute them as quickly as possible, then the children were to get dressed and go home. This took perhaps ten or fifteen minutes. But by the time the kids from the four classrooms were ready to go out, the storm had grown much worse. The teachers realized that they could not, that they must not, send the children out in these conditions. The youngest were hardly more than toddlers; the oldest were only fourteen. Somehow these excited, terrified kids had to be controlled and kept in school until it was safe for someone to go for help.

Visibility was so poor that none of them saw the drays approaching—wooden platforms mounted on bobsleds and dragged by horses, the nineteenth-century winter version of a flatbed truck. Five drays in all, each one attended by two men and drawn by two horses. They drove through the snow in a kind of ghostly procession, one dray right behind the other so that none would stray from the road and get lost.

For the teachers at their wits' end, the drays were a godsend. With transport and men to help, it was just a matter of getting the kids lined up and counted and then marched out the door and onto the drays. Again, Walter and the other monitors were called on to get their rows ready. The monitors must go last, after the other kids in their rows had filed out, one by one.

The air, when they finally got outside, was a shock. The air itself seemed to be streaming sideways in billows of grit. The snow felt like frozen sand against their eyelids and nostrils and lips. They couldn't face into the wind or open their eyes, even for a second. The wind was blowing so hard that if you fell you couldn't get up again. But to the kids it didn't matter. Being out in a storm powerful enough to shut down school and bring ten men out from town to rescue them was a tremendous lark, and the children fairly poured outside and down the rickety schoolhouse steps, everybody shouting over the wind and shoving and edging sideways or backward toward the drays.

The wood of the drays was already rimed with snow and frozen solid as rock, but for the first couple of minutes none of them felt the cold through their thin clothing. They piled on in masses of bodies and they had each other for warmth and a bit of shelter. There was much gleeful screaming as the schoolhouse emptied.

Walter took his responsibility as monitor very seriously. Not until his entire row was accounted for, assembled, and marched outside would he even dream of leaving the school. So he was one of the last ones out. The drays were nearly full by now—there was just room for him at the back of one. Walter scrambled up, the teachers did a final head count and shouted to the drivers that it was all right to start. The men snapped the reins and the five drays began creaking forward, one after the other in the storm, just as they had come out from town.

They hadn't gone ten yards when Walter suddenly hopped off. He had just remembered his precious water bottle. He knew enough about weather to realize that the water inside the fragile perfume bottle would freeze as soon as the schoolhouse stove went cold and then the ice trapped inside would burst the bottle. Without thinking, Walter dropped from the dray and rushed up the wooden steps, down the hall to his room, grabbed the bottle from his desk, and ran back out.

Only then did his thoughts catch up with his body. The drays had been barely creeping when he jumped off. He had assumed that they would still be in front of the school when he got back—or at least close enough to run after and overtake. Ordinarily, he could see for miles out here. Surely someone would spot him standing there and stop the dray and wait for him.

But that's not how it worked out. In the seconds that it took Walter to get his bottle, the drays had vanished without a trace— out of sight in the whiteout, out of earshot in the screaming wind. "The world is full of nothing" ran inanely through Walter's mind. And now he experienced that little seizure that tightens around the heart when you first realize you've taken a step that you cannot reverse. Snow clogged his nostrils and coated his eyelashes. Snow blew down the neck of his coat and up his sleeves. The air was so full of powdered ice crystals and it was moving so fast that Walter had trouble filling his lungs. The exposed skin of his face and neck felt seared, as if the wind carried fire not ice. A cottony numbness spread through his body and brain. It did not occur to Walter that he could still take shelter in the schoolhouse. Though he could barely see or breathe, he decided to set out for home.

Once he had made that decision, a door shut behind him. After a dozen steps into the storm, he could not have returned to the schoolhouse if he wanted to.

Countless witnesses wrote that visibility was so poor at the height of the blizzard that you couldn't see your hand in front of your face. It's tempting to dismiss this as hyperbole or a figure of speech—but there is in fact a meteorological basis for these claims. Dr. Louis Uccellini notes that the smaller the particles of the ice crystals, the worse visibility becomes—and there are numerous accounts of how the snow that day was as fine-grained as flour or sand. "There is a phrase used in blizzards of 'zero/zero' visibility," says Dr. Uccellini, "which means you can't see up or horizontally." This would explain how a woman near Sioux Falls froze to death with her key in her hand just steps from her door. And the husband and wife who both perished while blindly circling each other in their farmyard. And David Fyffe, the crusty Scottish cattleman in southwestern Minnesota, who was only able to traverse the 101 feet from his barn to his house (he was very precise about the distance) because by pure luck he stumbled on one of the bobsleds he tied up at the iron pump that was exactly halfway between. Had it not been for that fortunate accident he "would never have been heard of," Fyffe wrote in his memoirs. "There was no house nearer than a mile or more." A young Norwegian farmer, trapped in his barn in Dakota when the storm blew up, described how he "focused his mind on where the house stood in line with the barn" and then set out into the wind with hands out in front of him. It was only when his fingers caught in his wife's clothesline that he knew where he was. A homesteader in Buffalo County, South Dakota, wrote of a neighbor who was staggering lost between barn and house when he tripped over a snowdrift and fell against the house. The thud of his body hitting the house was loud enough to be heard over the storm and brought the family out to rescue him.

So it's hardly surprising that eight-year-old Walter Allen became confused and disoriented when he rushed out of the school clutching his perfume bottle and found himself alone in the storm. Even worse than the whiteout was the agony of his eyes when he tried to see through the snow. The fine hard pellets blew into his eyes and made them water. Walter cried and the snow mixed with his tears until it formed a crust between the upper and lower lids. Instinctively he reached up to brush the crust away with the back of his hand. Soon his eyeballs were inflamed, which further distorted his vision. The pain became so acute that it felt better to let the ice crust build. Tears and blowing snow melded together and sealed his eyes shut tightly. There was no way to break the seal except by tearing the tender skin.

Once Walter's eyes were gone, the rest of his face went fast. A mask of ice covered the exposed skin of his face except for holes at the nostrils and mouth. Snow penetrated his clothing and froze into an armor of ice around his body. All of this happened in moments.

Walter stumbled. The sizzle of driving snow hummed in his ears, and the frozen needles cut his face and throat. He knew he was lost. It was probably only a matter of minutes before he collapsed, whether in a gust of wind or because his feet became too frozen to bear his weight or from simple exhaustion we'll never know—and Walter himself didn't know either.

Strangely, once he was down, everything was better. On the ground the snow was softer and the wind didn't blow so hard. Walter curled up in the snow and surrendered.

The south wind had been in Johann Albrecht's face as he walked across the rutted fields to the schoolhouse, though it was soft for a winter wind and carried a smell of something damp and foggy. A fine January day—which only made his mother's tears and pleas more baffling. Peter, his younger brother, had given in, so he would spend the day at home listening to their mother cry and looking after Anna and the two baby brothers. But Johann was glad to be going to school—the English school, as his parents called it. No matter how dull the lessons or how repetitive the eternal chanting, school was better than farm chores. Two recesses a day, which was more than he got at home working for his father. With any luck the Graber and the Kaufmann boys would be there, too, and they'd have enough strong arms for a proper snowball war. The snow-balling got fierce when the Schweizer parents of Rosefield Township let all their sons attend school.

For a thirteen-year-old boy like Johann Albrecht, who had walked these southern Dakota fields all his life, the prairie didn't offer much to look at, especially in winter. To the west, the land heaved slightly so that a low rolling ridge blocked the horizon. The vista to the south went on forever. The plowed farm fields were either deep in drifts or crusted furrows, according to the whims of the wind. The saplings that his father and the other Schweizer farmers had planted in their timber claims—cotton-woods given to them by the government, mulberry trees, ash, elm, hackberry—were like crooked poles rising out of the snow. Every-thing else was sky—sky that seemed to revolve around you in slow circles when you walked out under it alone. Somehow you always felt a little foolish about singing or talking to yourself for fear somebody was watching or listening—though, of course, that was even more foolish, for who on earth could see or hear you out there except God?

There would have been but two columns of smoke in view—one from the neighboring farmhouse and one from the school, both of them curling slowly toward Johann as he walked into the southerly wind.

The school, like the Salem Church that the Schweizer families had built a few years earlier, was the simplest, starkest building possible—four rectangles capped by two triangles and a roof laid over the top. The door opened under one of the triangles on the west side and inside the door there was a tiny vestibule where the children hung their wraps on hooks. Two steps and you were standing in the classroom—a stove in the middle, crude wooden desks for the children, a dull light glowing at the small windows that had been cut into the long north and south walls. And at the front of the room, Mr. James P. Cotton, the American teacher who boarded with Johann Albrecht's family. It never ceased to amaze Johann how Mr. Cotton always seemed so much larger and more important in the classroom than he did at their house. This was his world, and here his word was law. Not a word of German could be spoken as soon as they crossed the threshold, though, of course, John—as he was called in school—and the others cheated sometimes.

Sure enough, when Johann reached the schoolhouse he saw that the Kaufmann boys and the Graber boys were already there—the other two families lived closer to school than the Albrechts; indeed, the Grabers were so close that the school was practically within shouting distance of their farmhouse. But surprisingly, they were the only other pupils in attendance that day. Seven boys altogether: strapping Johann Kaufmann and his two younger brothers, Heinrich and Elias Kaufmann; Peter and Johann Graber, sixteen and fourteen, and their eleven-year-old half brother, Andreas—three of the fourteen children who filled the Graber house.

And to round out the number to seven, Johann Albrecht himself, still known in the little Schweizer settlement as the baby who had been born on board the steamship *City of Richmond* during the great migration of 1874.

When Mr. Cotton asked Johann why his brother Peter was not

with him, Johann had to explain shamefacedly about the bad feeling his mother, Maria, had that morning and how soon after Mr. Cotton had left the Albrecht house for school, Maria had insisted that the boys stay home. Peter, weak of will, had listened to their mother, so Johann came alone, with his father's permission. School was too important to miss.

Mr. Cotton nodded his head. What was there to say? They had some strange ways, these Russian Germans. But at least they raised their kids to obey and to show respect; Mr. Cotton would grant them that much.

When the blast rocked the north wall of the school at around 11:00 in the morning, the boys and their teacher all turned to look at the north windows as if they had been summoned by a trumpet call. As everywhere, the wind and the darkness came almost simultaneously. The school windows went from pearl to charcoal as the cloud of snow enveloped them, not so much falling as slamming sideways. Within minutes the wind had sucked the warm air out of the uninsulated building. Powdery snow began sifting in through every crack in the walls and around the window frames and spraying against their faces. Soon there was fine snow hanging in fringes from the maps on the wall and eddies of snow snaking across the floorboards. Even a few feet away from the stove, it was so cold that the snow didn't melt. White cobwebby drifts mounted in the corners.

Instinctively, the younger boys looked to their older brothers to see what to do. Heinrich and Elias Kaufmann knew that Johann would take care of them. Their kindly mother, Anna, always told them that if anything happened, they must listen to Johann. He and Peter Graber were practically men. They would know what to do in a storm like this. Probably better than an outsider like Mr. Cotton.

Johann Albrecht, with no brother at school, turned to Mr. Cotton. But the teacher would not meet his eyes. He just kept looking over at the rattling windows and shaking his head. With the wind roaring in their ears and the room going colder every minute, there was no question of trying to teach any longer. But what should they do? Usually he gave the commands and the boys obeyed, no questions asked. But now Mr. Cotton was asking the two older boys for advice and then arguing with them. Johann Kaufmann and Peter Graber drew together. They must stay in school and wait for their fathers to come for them. That was what they had always been told to do in a blizzard. Keep under shelter. Remain where you were. Stay together.

But no. Mr. Cotton had made up his mind. They must leave the school and go to the nearest house, where they would find food and warmth. All of them would go together to the Grabers' place—less than a quarter of a mile away, a few hundred paces. They would take shelter with the Graber family and wait out the storm there.

Mr. Cotton ordered them to get their wraps from the little vestibule and get ready. The Schweizer boys looked at each other with long faces and shook their heads, but they obeyed as they had been raised to. The two older boys went outside first, each one wincing as if slapped when he stepped into the wind. When all eight of them were outside in the storm, they huddled for a minute behind the south wall of the schoolhouse, which was the only place they could stand up straight. But even there in the lee of the building they had to hunch their shoulders and drop their faces against the onslaught of tiny crystals. Then they set out, one by one, into the wind.

As soon as they were outside, Mr. Cotton's authority vanished as if it had been torn from him by the wind. Within a few paces, the boys had split into two groups. Peter Graber and Johann Kaufmann, the two sixteen-year-olds, went first and three of the younger boys went with them. Heinrich and Elias Kaufmann stayed close to their brother Johann, as they knew their mother,

Anna, would want them to. Johann would take care of them and see that they got home safely.

Johann Albrecht fell in with this group, too. He was thirteen, almost as old as Johann Kaufmann and Peter Graber. They didn't need the teacher to find their way to the Grabers' house. They weren't babies.

Andreas and Johann Graber got separated from their older brother, Peter, and ended up with Mr. Cotton. Later Andreas would have a hard time explaining what happened. It all went so fast and it was so difficult to see anything with the wind lashing needles of ice in his eyes. They had started out together, staggering one after the other into the wind, but then their brother Peter and the Kaufmann boys and Johann Albrecht somehow got in front. One moment they were there, ghostly shapes a few paces ahead. The next moment they were gone and there was nothing where they had been but the stinging white.

The drifts were already too deep for Andreas to walk through. Andreas began to flounder in the snow and his hands ached with cold. He couldn't see Mr. Cotton or his brother Johann anymore. They were going on without him. He couldn't see anybody. Andreas panicked and shouted for Peter to come and help him. But instead of Peter, Mr. Cotton and Johann appeared. As they loomed out of the snow, Andreas heard Mr. Cotton calling ahead to the five other boys to wait for them. But there was no answer that Andreas could hear over the wind.

Mr. Cotton told Andreas and Johann that the other boys must already be at the Graber house. Peter would know the way home. He must have led the others there. The house couldn't be far. The five boys would be there waiting for them.

So they staggered on—Mr. Cotton in the lead and the two younger Graber brothers behind. Strangely, Andreas did not feel the cold anymore. In fact, he felt a kind of warm glow spread through him. So he didn't cry when Mr. Cotton stopped and turned around and shouted that they must have missed the farm-house. He wasn't scared when his brother Johann wailed in German that they were lost. Maybe they could make it back to the school—but which way? If they had passed their house, then what? Where was the next house? Or even a barn? Mr. Cotton turned slowly to peer in every direction. Nothing.

※ ※

It was the row of spindly trees that saved them. Andreas's father had planted the trees soon after he settled in Dakota. Most of the Schweizer farmers did the same—an orchard of fruit and nut trees near the house, just like the ones they had in the Ukraine. For an instant, the air cleared enough for them to see the end of a row of his father's fruit trees. Andreas now knew exactly where they were. The row began at the house and ran due east. All they had to do was follow the line of saplings back to the house. It was the kind of lucky stroke people always called a miracle. Had the storm not abated just at that moment, they never would have seen the trees, never would have realized that the storm had forced them well east of where they meant to be. Had it not been for the trees, they would have continued drifting east with the storm until the wind finally blew them to the ground. Luck. Pure luck.

As soon as they got inside the farmhouse Mr. Cotton asked the Graber parents, "Are the other boys here yet?" Andreas's mother and father shook their heads—they didn't understand. So Johann shouted the same question in German. Peter and Susanna Graber stared wide-eyed at the two boys. Andreas knew it was a foolish question. This was no mansion with rooms to hide in. If their brother Peter and the Kaufmann boys and Johann Albrecht had made it to the house, there would be no need to ask.

※ ※

Those who were fortunate enough to be inside when the storm came up faced the same dilemma as passengers on a ship who have just seen one of their fellows fall overboard. To stay indoors and do nothing seemed heartless, but to venture forth on a rescue mission was likely to be fatal to the rescuer and useless to the lost. What would be the good of another soul wandering blind in the storm? So the settlers kept candles burning through the night at their windows or they stood at their doors shouting or ringing bells or beating on kettles or washtubs with hammers or spoons. They strained to hear cries for help over the wind. They prayed.

They prayed for the nineteen-year-old teacher Etta Shattuck in the house where she boarded in Holt County. Etta had made her plans clear before she left the house late in the morning. She had already closed her school. Tomorrow she was heading back to Seward to rejoin her family. Once she collected her final wages, she would go. That was why Etta had walked out that morning—to get her order signed so she could be paid her twenty-five dollars. If it hadn't been for that, Etta would have been inside and safe when the blizzard struck.

The man at whose house Etta boarded stood at his door and shouted for as long as his lungs held out. He knew which way she had walked—he had seen her disappear. But there was no question of going after her. Didn't he have a wife and children of his own to think about? Etta was a strong and sensible girl, very settled for nineteen, religious, too. She knew every hymn in the hymnal, it seemed. When it got darker and colder and she still didn't come back, he and his wife prayed that Etta had found shelter. That was the best they could hope for.

But that's not what happened to Etta Shattuck.

Etta was not far from the house where she boarded when the storm blew up. There was a forty-acre pasture around the house that the farmer had fenced in, and Etta was still inside the fence-line. That was lucky, for fences were rare in that part of the country. Eventually, if she kept going in a straight line, Etta would hit the fence. And then, no matter which way she turned, the fence would lead her back to the house. All she had to do was hold fast to the fence and follow it around the pasture. If she kept her wits about her, the fence would save her.

Etta knew this, and when she reached the fence she followed it as best she could. But then she began to doubt. It was hard to think straight with the wind slamming against her head. She had been following the fence for what felt like miles and still no sign of the house. Surely she had come too far. She began to be confused, and confusion made her desperate. The fence was not guiding her but trapping her.

In her bewilderment and rising panic, Etta made a move she could not reverse. She bent over, put her hands and knees on the snow, and crawled under the fence. Two steps and the fence was gone as if it had been erased. Now she was in open country with no barrier or landmark for miles. The odds of stumbling upon a farmhouse or sod hut in zero visibility were essentially nil.

At the mercy of the storm, Etta drifted with the wind and prayed. The habit of prayer had taken hold strongly since she became a Christian at the age of sixteen. She prayed and sang hymns. Etta knew God had a reason for unleashing the blizzard. God, who brought the storm, would guide her steps to safety. God would not let her die on the prairie. She prayed to God to give her shelter.

And sure enough. A haystack suddenly loomed before her.

By the time she reached it, Etta was nearly giddy with fear and exhaustion. She was too weak to dig in very much. She had no pitchfork and her hands were frozen. Etta managed to scrape off some of the hay and make a cavity for herself. But it was shallow. When she fell into the hay, the cold was still biting at her legs. She pulled her legs up and covered them the best she could. She knew she needed to burrow in deeper but she couldn't. Her hands didn't work, the fingers wouldn't close. She couldn't summon the strength to get up again.

Etta prayed to God to watch over her.

In Huron, 130 miles due north of where Etta Shattuck lay in a haystack, Sergeant Glenn was now jotting down the wind velocity every few minutes—11:45 A.M. (Central time): 42 miles per hour; 11:47 A.M.: 48 miles per hour; 12:15 P.M.: 57 miles per hour from the northwest; 1:30 P.M.: a maximum sustained speed of 60 miles per hour. He estimated the gusts at 80.

Five minutes after the wind reached 60 miles an hour, Glenn received Woodruff's telegraph to hoist the cold wave flag. There must have been a problem on the Huron line or a backup at the local Western Union office, because most of the Nebraska and Dakota observers had received their telegrams an hour earlier—12:30 P.M. Central time in Crete, Nebraska; 12:20 at North Platte; 12:20 at Bismarck, where a gale had been blowing since 5:30 in the morning. In Omaha, Sergeant George M. Chappel, the Signal Corps observer in charge, wrote in the station journal that the arrival of the cold wave warning at 12:40 P.M. was "not far enough in advance of the cold wave to enable this office to get the warning telegraphed to the northern and western portion of the state before the blizzard had struck there."

"Not far enough in advance" was putting it mildly. In Moorhead, in far northwestern Minnesota, fifteen minutes after Woodruff's telegram arrived at 12:30 P.M., Private Frank L. Harrod noted a "sudden and fierce change of wind from south to north" followed by "heavy blinding snow."

On a map the advance of the cold front looked like the lobes of a glacier advancing out of the northwest. By 1 P.M., it had covered almost all of Dakota Territory, the western two-thirds of Nebraska, and the northwestern fringe of Minnesota. Over the next two hours the front picked up speed as it spread inexorably over the most populated section of the prairie. Only three Dakota Signal Service stations had failed to hoist their cold wave flags by now: Huron and Yankton, where Woodruff's message arrived late, and Rapid City, way out in the Black Hills, where the telegraph lines were down and the message didn't arrive until the following afternoon. But it made little difference. Nobody knew about the flags except those who lived within sight of the stations—and even then, visibility was so bad as to render the black square at the center of the white flag invisible unless you were a few feet away. In all the hundreds of accounts written by those who endured the blizzard, there is scarcely a mention of the appearance of the cold wave flag. Though Greely boasted that the system of cold wave warnings operated "to the general satisfaction and frequently great advantage of the public," the public never breathed a word of gratitude—or complaint. Whatever Greely believed, the people of the region knew they were on their own when a blizzard hit.

Particularly the schoolteachers, many of them barely older or more educated than the children they taught. As the blizzard broke against the northwest walls of their schoolhouses, every teacher faced the same choice alone: stay in the school with the children or send them home? According to Sergeant Glenn, those who chose the former were "principally persons familiar with the western storms and who fully appreciated the danger of going over the open prairie in a 'blizzard.'" But not all of those who left or dismissed the children were ignorant or careless. Some took the children out into the storm only after their fuel was "exhausted," wrote Glenn, and "it became a question of freezing where they were or in the attempt to find other shelter, more comfortable; others started out supposing they could go in safety but were soon bewildered and lost."

Minnie Mae Freeman was one of the many teenage teachers who faced the question of freezing or fleeing in the storm. She had sixteen pupils, some nearly as old as she was, at her country schoolhouse near Ord in the Loup River region just east of the Nebraska Sand Hills. It was a sod building, unusual for a schoolhouse on the prairie, with a crude door attached by leather hinges and a roof of tar paper with sod laid over it. Around noon, the first blast of the storm tore the door off the leather hinges and blew it into the schoolroom. A couple of the boys helped Minnie get the door back up, whereupon it blew in again. This time she had them nail it shut all around. Minnie knew she had enough coal to heat the soddie schoolhouse all night, and she was determined to stay put and keep the children inside.

When a gust ripped off a piece of tar paper at the top of the roof where the sod had fallen away, Minnie realized that they would all eventually die of cold if they tried to stay. The family with whom she boarded lived half a mile north of the school and she decided that the best plan was to take the children there for the night. According to some accounts, she found a length of rope and tied the children one to another before setting out. Others insist the roping up was pure myth fabricated later to glorify the teacher. A student named Emma Lee, who was present that day, wrote that Minnie had the children crawl out through the south window since the door was nailed shut. "The nearest house was not quite a quarter of a mile away," recalled Emma. "We could have gone there with the storm at our backs. However we were told to stay with the teacher and go to her boarding place, which was a half mile away, and we had to face the storm. We were not tied together in any way, as has been erroneously stated so many times." Several of the smaller students stumbled and fell on the way, and at one point Minnie fell.

Minnie Freeman always insisted that she deserved no special attention—that she had done what anyone would do. But in fact many older and more experienced teachers failed to act as quickly and as sensibly. All of Minnie's students stayed together and made it safely to the two-room sod house where she boarded.

Stella Badger, another young Nebraska schoolteacher, had a one-room schoolhouse in a hilly area south of Seward, about a hundred miles southeast of Minnie Freeman's school. This was the stretch of land they called the Bohemian Alps—the rough, broken, ravine-choked sections that more prosperous or canny farmers had sold to German and Czech immigrants. A good blow in winter would fill the ravines between the hills with snow to depths of ten or fifteen feet.

Stella Badger's school sat on a rise just south and west of a web of deep ravines with creeks running through their bottoms. Though the local children could have walked to and from school along the straight section roads, they always cut through the ravines. Miss Badger told them again and again to take the roads, but the German and Czech kids had a way of not understanding what they didn't want to hear.

Lena Woebbecke was one of the German kids who never seemed to understand. Miss Badger couldn't tell if she was backward or hard of hearing or just headstrong. Lena wouldn't meet her eyes when she talked to her. She kept to herself. Large for eleven, scarred from smallpox, the child had just started at school in the fall. Miss Badger heard she'd been sent to live with the Woebbeckes because her own mother had abandoned her. All through the autumn Lena barely spoke, even to the other German children.

When the wind slammed against the north side of the school-house around three o'clock on the afternoon of the twelfth and the classroom went suddenly dark, Miss Badger ordered the children to stay just where they were. She would not dismiss school until the storm ended. The ravines would be too dangerous already and even the road would be hard to see in the blowing snow. She told them as calmly as she could that there was enough coal in the coal house

to keep them going all night if need be. It was a relief that none of
the children put up a fight.

Miss Badger surveyed the faces before her and drew a mental
map of where each child lived. The smaller children and the girls
worried her most—but nearly all of them lived south of the school
near the place where she boarded. The terrain wasn't as rough in
that direction. Only one of the girls lived north of the schoolhouse,
beyond the ravines on a deeply cut up section of the Bohemian
Alps. That was Lena.

Miss Badger was wondering whether they should all try to make
a dash for her boarding place when she heard the sound of boots
stomping in the little vestibule. The door to the classroom opened
and a farmer who lived nearby appeared in a gust of cold and snow.
He looked like he'd just dug out of an avalanche. Snow clung like
lint to every bit of his clothing, to his eyebrows and hair. The
farmer shouted over the wind to Miss Badger that he had come to
take his children home. Their farm lay south of the school so they'd
have the storm at their backs on the way home.

Miss Badger and the man talked for a few minutes and the two
of them looked the children over. All of the pupils but two—Lena
and an older boy—lived south of the school, as did Miss Badger
herself. The farmer promised that he'd see all of them home safely.
He had brought blankets for the smaller children. It would be
madness to spend the night in the classroom with no food. Even
with the coal fire in the small stove, the drafty frame building was
getting colder by the minute.

But what about Lena? Miss Badger knew the German girl would
never agree to accompany her to her boarding place or spend the
night with one of the other children. Lena was stubborn. She would
insist on getting back to the Woebbeckes. But how could Miss
Badger send one girl out alone in the storm?

The older boy who lived near the Woebbeckes came to her rescue. "I can see to her," he said, pointing to Lena.

And so it was settled. Miss Badger and the farmer would take all the children who lived south of the school. Lena and the older boy would go together the other way. The farmer was urging Miss Badger to get moving before the storm became worse. There really wasn't time to think through the plan.

The schoolhouse was set back just a few feet from the dirt road that ran along the section line and that's where the two groups parted. Miss Badger was the last to disappear, her dark skirt wrapped around her legs, her body tilted back at a strange angle from her heels so that the wind didn't blow her over onto her face.

As soon as Lena and the boy were left alone they began to argue— or rather the boy argued and Lena pulled her thin cloak around her shoulders and looked at the ground. The wind was full in their faces on this exposed hillside. In the few minutes they stood there on the road Lena felt the floury snow work its way down her collar and up the cuffs of her sleeves. The air was moving so fast and was so dense with crystals she could hardly breathe. The boy screamed over the wind that he was taking the road—and he pointed down the hill. Lena stood her ground and shook her head. She would cut across the stubble field behind the school and take the ravine as she always did. The boy hollered and pointed again. And then he became exasperated and gave up. She could suit herself. He started down the hill on the straight road and instantly vanished. Within an hour he was safe at home.

Being alone did not make anything worse for Lena—at first. She knew the way. She didn't need the boy. She tightened her hands around the lunch pail and the reader that the Woebbeckes had given her and made her promise never to lose or leave at school, and she walked down the path she always took home. She could feel the ground descending beneath her feet, but with the snow frozen to her eyelashes she couldn't see anything. Lena was about halfway home when she stopped. She knew instinctively that she had wandered off the usual path.

Lost. The fact is usually irrevocable long before the brain produces the word. Lena was lost. She was seized by that rising panic when the chest goes hollow and the heart races in the ears and the body slicks itself in sweat. But her stolid temperament served her well. Lena was not flighty or hysterical. Ever since her father died she had looked after herself. Lena kept her wits about her. With her freezing fists still wrapped around the lunch pail and reader, she sensibly turned around and attempted to retrace her steps. Back to the school. Hadn't Miss Badger said there was plenty of coal?

Lena made it to within a hundred yards of the district 71 schoolhouse and then her strength gave out. She collapsed in a furrow of the stubble field adjoining the school. Her last conscious act was to cover herself with her cloak.

Every object prominent enough to catch the wind already had a drift shadowing its lee. Haystacks, fence posts, houses, barns. When she fell on the side of the hill below the schoolhouse, Lena became one of those objects.

Signal Corps observations taken simultaneously at 2 P.M. Central time:

Fort Assinniboine, Montana: Pressure 27.49 and rising; temperature 22 below zero, north wind at 9 miles per hour.

North Platte, Nebraska: Pressure 26.79; temperature 2 below zero (a fall of thirty degrees in the past eight hours); north wind 40 miles per hour.

Huron, Dakota Territory: Pressure 28.29 and rising; temperature two below zero (a fall of twenty-one degrees in the past eight hours); northwest wind 44 miles per hour.

Yankton, Dakota Territory: Pressure 28.16 and falling; temperature eighteen above zero; northwest wind 18 miles an hour.

Omaha: Pressure 28.31 and falling; temperature 27 degrees above zero; southeast wind 9 miles per hour.

The cold front was now so well defined that it looked like a giant comma dropping down from the northwest. In the course of the day the upper part of the comma had elongated and assumed a nearly vertical north-south orientation so that by 2 P.M. Central time, when the observers checked their thermometers and barometers and anemometers, the front sliced a clean line down the Minnesota-South Dakota border and then bent slightly southwest to lop the eastern triangle of Nebraska off from the rest of the state. Huron was already behind the front; Omaha would fall within the hour; Saint Paul not until after sunset.

At the two o'clock observations the Rollag family farms in the extreme southwestern corner of Minnesota and just over the border in southern Dakota exactly straddled the leading edge of the front. Austin Rollag, his mother, Kari, Ann, his wife of five years, and their small children lived on the Dakota side. His sister, Gro, now known as Grace, and her husband, Ole Rollag, lived with their six children just east of the Minnesota state line in Beaver Creek. The storm hit the two farms within a matter of minutes, Austin's place first, then Ole's.

Grace Rollag thanked God that at last she had a frame house after enduring the soddie for seven years. And a proper barn for the animals. It would have been impossible to live much longer in the sod hut with six children. Six American children, all born in Minnesota. Peter or Peder, Charley or Carl—it didn't matter what they called themselves, they were Norwegian underneath. And of course the older they got, the more Norwegian they looked—Peter now thirteen and Charley ten, with their clear faces and high brows.

Grace was alone at home with her four younger children—Nels, Clara, Susan, and Anna Marie—when the house went dark and the wind began to roar so loudly she couldn't hear the children's voices. Ole and the two older boys were outside working—Peter watering the cattle at a spring about two hundred yards north of the house, Ole and Charley bringing back a load of hay from a pasture in the bottomland about a mile southeast of the house. Grace had been on the prairie long enough to know what that roaring meant. She didn't pause to think. In a matter of seconds she had pulled on a pair of spare boots, wrapped herself in a cloak and scarf, and told Nels, who was nine, that he must look after the baby. Her oldest son, Peter, was in trouble. Even with the wind at his back, Peter at thirteen was too small and too young to drive the cattle from the spring to the barn and get them all under shelter by himself. She must go out to find Peter and help him. As for Charley, he had his father to look after him. Ole would get the boy to safety—with God's help. Grace must see to Peter.

Before she opened the door and put her head down in the wind, Grace paused to collect her thoughts. She knew that once she got outside it would be impossible to see or to think and that even the smallest error in judgment could be fatal—for her, for her son. House, barn, spring: Grace fixed a picture in her mind of the layout of their property and then she set out to find her son.

※ ※

Ole and Charley had gotten the wagon loaded up with hay and were about half a mile from home when the storm overtook them. The wind came up so fast that it knocked the wagon over and spooked the horses. Ole shouted to the boy that they must unhitch the horses and get them home. The hay could wait. They would come back for the load after the storm. The thing to do was calm the horses down and find shelter.

Ole Rollag was a big, burly, bearded man who was accustomed to imposing his will on the world around him by brute force—a true pioneer—so he figured on muscling his way through the storm. He'd grab one horse, the boy would take the other. Ten was plenty old enough to lend a hand.

The problem was that they'd been working southeast of the house. To get back they would have to walk straight into the wind. Even if Ole and Charley could have staggered that distance, the horses refused. Try as they might, they could not get the horses to walk with their faces into the wind. Buffalo will stand stock-still in a blizzard with their heads down and their thick hides going gray with snow and wait for the wind to blow itself out, but horses and cows tend to drift where the wind takes them, sometimes for miles, before they die of exhaustion or freeze to death or break their legs in gullies. Ole and Charley fought the horses for all they were worth, but the best they could do was to turn them northeast. Ole was well aware that this wouldn't get them home—but it was better than doing nothing or losing the horses altogether.

Austin and his wife, Ann, were also caught outside in the blizzard. Eager to use every minute of the good weather that morning, they had left the children at home in the care of Austin's mother, Kari, and gone out to work in the barn. Since the morning was mild, they had turned their animals—a few horses, a foal, and some cows—into the pasture near the barn, while the two of them worked close by. "After noon, about 3:30, we heard a hideous roar in the air," remembered Austin. "At first we thought that it was the Omaha train which had been blocked and was trying to open the track." When he looked up at the sky, Austin saw the snow descend "as if it had slid out of a sack. A hurricane-like wind blew, so that the snow drifted high in the air, and it became terribly cold. Within a few minutes it was as dark as a cellar, and one could not see one's hand in front of one's face."

Like almost every farmer caught out that day, Austin and

Ann's first thought was the safety of their animals. Never mind that they had young children and an aged parent at home—they had to get the animals under shelter. The stock had not strayed far from the barn so it was relatively easy to drive them in. All except for a foal, which had gotten separated from the other horses. There was no way a horse that young could survive a prairie blizzard. Austin and Ann wouldn't leave the farmyard until they had found the foal.

Old-timers on the prairie will tell you that a lull often follows the first blast of a blizzard, a short pause before the storm really lets loose, and meteorologists confirm this observation. It's like that sharp intake of breath between a baby's initial scream and a full-fledged tantrum. The lull was what cleared the air enough for the younger Graber boys and Mr. Cotton to see the row of saplings and find their way to the farmhouse. This same lull saved the life of the Rollags' foal—and probably their own. In that brief moment of calm Austin and Ann heard the foal whinny. They heard him and then they saw him. Now all the animals were safe.

Only after the foal was shut up in the barn with the rest of the animals did it occur to Austin and Ann that they couldn't see their house. Pioneer families always argued about where to put the barn: too close to the house and you'd get flies and the odor of the animals all summer; too far and you'd waste time walking back and forth and risk your life in storms like this one. Austin and Ann now realized that they had erred on the side of too far.

It was Austin's mother, Kari, who saved them. When she heard the storm come up, Kari somehow found an old cowbell inside the house and she grabbed it and stood just inside the door and rang it into the wind with everything she had. Austin and Ann heard the faint clangs over the keening of the wind and followed it home like a beacon.

Grace Rollag may not have been the best housekeeper in Minnesota—she was too fond of reading to stay on top of every little chore—but she knew how to handle herself when life got rough. She was not one to get lost in a blizzard.

Grace drew up her shawl as she stepped out of her house, bowed her head into the wind, and marched straight ahead to where she reckoned the barn was hidden behind the blowing snow. Fifteen years of farmwork and childbearing had thickened Grace's body and coarsened her hands. The waist she was so proud of in the photograph taken on her wedding day in April 1873 was long gone and she'd never again put on that black dress with the white lace collar. Never mind. A willowy waist would not help her now.

Grace reached the barn with no problem. By sheer luck Peter had found his way there just moments before. As Grace had predicted, he'd gotten the cattle up from the spring, but now he was having a hard time. It was the door to the barn—the snow had already drifted so high that Peter couldn't get it open. Grace tried her best, but it was hopeless. Even if they shoveled it out fast enough to open it, they'd never get the door shut again. They'd have to drive the cattle around back by the horses' stable and get them into the barn that way. The door here was sheltered from the wind, but it was narrow and gave on a tight passage between stalls—just wide enough for a single cow. Cows were always balky about entering tight spots, but there was no other choice.

Somehow, with sticks and shouts and desperate brute force, Grace and her son forced the cows into the door to the horse stable and led them one by one down the passage through the stable and into the barn. Every last one was safe. Even in the cold Grace and Peter were sweating by the time they got the job done.

Luck was also with Ole and Charley that afternoon. Blinded by the snow and staggering at the mercy of the wind, they might have ended up anywhere. By dead chance the wind deposited them at the neighboring farm, about a mile to the east. At first Ole couldn't believe it since he was convinced they had been traveling *northeast*. But the northwest wind was so powerful that they had barely gone north at all.

Once he had gotten out of his frozen clothes and thawed out enough to see straight, Ole was shocked by the condition of Charley's face. The fair skin of the boy's cheeks and forehead, which usually turned blotchy red in the cold, was cheesy white with frostbite. Ole knew the boy would also feel worse as he thawed.

The neighbor's wife took Charley in hand. She rubbed the frozen skin with snow so it would thaw gradually. She'd see that there would be no scars.

When the boy was settled and a bit of color returned to his face, Ole made up his mind. Grace would be frantic if he didn't get home that night—so frantic she might go out looking for them. She might be out looking now. It was a mile to home—a mile due west, not directly into the wind, but bad enough. No matter. He wouldn't be wrestling with the horses—they could stay in the neighbors' barn until the storm was over. And there was no question of taking Charley back out again. He would spend the night here. So Ole would be walking unencumbered. It would be easier this way. He would walk the mile to his home and find Grace and the other children. Before it got any darker, Ole set out by himself into the storm once more.

❧ ❧

The blizzard reached Lincoln, Nebraska's capital, at 3 P.M. Snow had been falling on the city since the early morning, a dreamy Christmas kind of snow with large wet flakes coming straight down through the white windless air. Then, at three o'clock, the sky suddenly blackened and the wind veered from south to north. Four minutes later, it was impossible to see even the outlines of buildings across the street. All the city streetcars were immediately taken in and teamsters rushed to haul drays and wagons to cover. A few last straggling hackneys battled the rapidly mounting drifts. Pedestrians ran for their lives.

At Lincoln's Capital Hotel the assembled committeemen of the Nebraska Press Association descended anxiously on the front desk. Their convention was scheduled to convene at the hotel at eight o'clock that night, though few of the members had arrived. Association president Bushnell had reason to believe that some 125 esteemed Nebraska newsmen were either en route from other parts of the state or had yet to set out. Among the former was the association's secretary, MacMurphy, who was at that very moment sitting on a stalled Lincoln-bound train enshrouded in sheets of snow. Bushnell inquired of the desk clerk whether he had any knowledge of when MacMurphy and his fellow Nebraska editors and reporters were likely to arrive at the Lincoln train station. Indeed he did. The clerk had it on the best authority that not a wheel was turning on any of the ten railroads that served the capital. Those unfortunate enough to be traveling were therefore stranded without hope. Those who had not yet set out would have to wait for the storm to abate, which at the moment it showed not the slightest indication of doing.

※ ※

Ten minutes after it rolled over Lincoln, the storm was upon Crete, nineteen miles to the southwest. Signal Corps observer Private C. D. Burnley noted in the station journal that after the wind shifted "the temperature fell 18 degrees in less than three minutes. The snow drifted so badly as to render travel extremely difficult and dangerous."

※ ※

An hour later the blast had reached Omaha. At the Signal Service station in the Custom House building at the corner of Dodger and First, Sergeant Chappel clocked the wind shift at 4:17 P.M. Central time. By 5:30 P.M. all streetcars in the city ceased to run because it was impossible to keep the tracks clear of drifts. No trains left Omaha that day.

The storm overtook a large party of local citizens who were out celebrating the completion of a new bridge over the Missouri River between Omaha and Council Bluffs on the Iowa side. Some four hundred sleighs and cutters had paraded down Omaha's Douglas Street earlier in the day and crossed to Iowa—not on the new bridge, but on the partially frozen river. The citizens' plan was to dine and dance in Council Bluffs, and then return to Omaha by starlight.

The bridge was narrow. There were gaping holes in the ice on the river. There was no safe way to cross back to Omaha. As the blizzard raged, an urgent telegram arrived at the office of the Omaha *Republican* from a reporter who was with the group in Council Bluffs: "Turn the whole force loose on possible loss on the river."

Farther east, in Des Moines, a parade was forming to honor the inauguration of Iowa's governor, William Larrabee, just elected to his second term. Company B, Second Regiment of Davenport, was supposed to escort the Governor's Guards from Kirkwood House to the State Capitol—but the troops were delayed in the storm and the procession left without them. The governor's carriage rolled slowly through the drifting snow with a straggling accompaniment of mounted officers from the Third Regiment. There were many empty seats in the Capitol rotunda as the Republican Governor Larrabee launched into a lengthy address attacking the railroads for encroaching on the rights of the people and arguing for more stringent regulation of large corporations.

In the middle of January, the sun sets a few minutes after five o'clock over the prairie states. Professor William Payne, at his observatory on the campus of Carleton College, would have known precisely how many minutes and seconds after five o'clock and would have seen to it that anyone who was curious knew the exact instant of sunset as well. Not that the official time really mattered on the afternoon of January 12. It was the advancing storm, not the sun, that controlled the intensity and duration of light that day. The blizzard created its own sunset, which moved southeast down the prairie at 60 to 70 miles an hour.

Sergeant Glenn reported that the storm hit Huron a few minutes before noon, so Wessington Springs, which is twenty-six miles to the southwest of Huron as the crow flies, must have gone dark just a few minutes later. From noon until 4 P.M., May Hunt did her best to carry on in the roaring twilight with the seven students who had come to the country school she taught near Wessington Springs. Four o'clock was when the fuel ran out and May Hunt and the seven children were suddenly faced with the choice, in Sergeant Glenn's words, "of freezing where they were or in the attempt to find other shelter, more comfortable."

May Hunt chose to go. Just 140 yards west of the school, on the other side of a gully, there was a farmhouse belonging to the Hinner family. The children in school that day—Fred and Charles Weeks, the three older children of the Reverend S. F. Huntley and his wife, Abi, and Frank and Addie Knieriem—all lived at least three-quarters of a mile away. The Hinner place looked like the best option.

The gully was what worried May. It was five feet or so to the bottom and the sides were steep. There was a culvert thrown across it that served as a bridge, but it might be hard to find in the storm. If they missed the culvert bridge the children would be in danger of falling in the gully—and in drifting snow the smaller ones would have trouble getting out again. May counted herself lucky that Fred Weeks had come to school that day. At eighteen, Fred was by far her oldest pupil, a big shy dark-haired farm boy with a round face and large hands. When May told him about her plan of taking the children over to the Hinner place, Fred volunteered to go scouting. If he could find the bridge, he'd clear a path and then come back for the others.

Fred was gone for half an hour while the rest of them stood around the dying fire. When he finally came back the younger children cheered. He told them that he had walked back and forth between the school and the bridge not once but twice. They'd be all right following him. Once he got them all across the bridge, they could pick up the path and take it to the Hinner house.

It was after 4:30 by the time all seven of May Hunt's students were ready to go. Addie Knieriem, one of the younger girls, was wearing thin, dainty little shoes, so Miss Hunt wound scarves around each shoe to keep her feet from freezing. More time was lost while Miss Hunt arranged her hat and veil. Fred went first, and when they got outside they all joined hands behind him in a human chain. Even in those few minutes the blizzard had gotten worse. The track that Fred had kicked into the snow was completely drifted over. There was no sign of the bridge that he had just crossed twice.

One hundred forty yards separated the Knieriem school, as it was called, from the closest house—nearly half again as long as a football field. On a clear day all but the youngest of the children could have walked it in ten minutes. But in the storm, blinded and deafened and barely able to breathe, the children quickly came to grief.

It was the gully that brought on panic. Stepping out where he hoped the culvert bridge would meet his shoe, Fred fell through the snow that had drifted into the gully and dragged the others down with him. They toppled over one after another like bowling pins. It would have been funny—certainly the children would have thought so—if they weren't so frightened. As they plunged, the teacher's veil tore away from her hat. She became aware that Addie Knieriem was in trouble. The scarves had come off Addie's shoes as the girl floundered up the far side of the gully. More precious time and body warmth were lost fussing with the teacher's veil and the child's shoes.

Somehow they all got up out of the gully and reassembled. Eight of them counting May Hunt. Again Fred went first. Since they had missed the bridge it was hopeless to try to find the path. But if the wind relented even for thirty seconds maybe they could make out the Hinner house. They knew it sat on a rise just above the gully—a stone's throw away. With every step they expected to catch a glimpse of the house through the gray horizontal snow. By now the sun had set and what little light remained was quickly draining out of the air.

It's hard to fathom how children who walked to and from school a half mile or more every day became exhausted to the point of collapse while walking a hundred yards that afternoon. Hard to fathom until you consider the state of their thin cotton clothing, their eyelashes webbed with ice and frozen shut, the ice plugs that formed inside their noses, the ice masks that hung on their faces. This was not a feathery sifting of gossamer powder. It was a frozen sandstorm. Cattle died standing up, died of suffocation before they froze solid.

When they got out of the gully, May Hunt's students were wet to the skin and nearly blind. Most had lost the use of their fingers. Addie Knieriem had no sensation in her toes. Panic wicked away what little heat remained inside their bodies. All of them were shivering violently and the younger children looked limp as rag dolls. These children, all but Fred and his brother, were already desperate and ready to give up. It happened that fast.

If you had your choice of what to burn to keep your house warm, the straw of the oilseed flax plant would surely be near the bottom of the list. Unless it's processed and compressed into briquettes, flax straw burns so fast that you need a small mountain of it to heat even a shack or sod house through a long prairie winter. It's messy and brittle and needs to be twisted or braided into bundles to burn efficiently. But in a place with no trees and not enough money to buy coal, flax straw or hay were better than freezing to death and cleaner than buffalo chips. After a snowstorm on the treeless prairie, the first set of tracks usually led from the farmhouse door to the mound of the straw pile, the second set to the barn.

Farmers generally kept their straw or hay piles just a few yards away from their houses, so when Fred Weeks stumbled into the Hinners' pile he figured they were saved. The party could take shelter in the straw pile for a few minutes while Fred went to look for the farmhouse. Then he'd come back and lead them to safety. The Hinners would give them food and let them warm themselves by the fire of twisted straw. That was how it was supposed to happen.

By a stroke of luck Fred found a pitchfork and a piece of lath next to the straw pile and he enlisted his brother, Charles, to help him dig out a cave in the sheltered side. The exertion warmed the boys and cleared their heads. It took some time to hollow out a space big enough for Miss Hunt and the children. When it was done they piled in and huddled together, panting and gasping. It was better, far better, than standing out in the vicious wind and snow.

Miss Hunt insisted that Fred must not go out looking for the farmhouse alone. Charles would go with him, and the Huntley boy, Ernest, volunteered to join the Weeks brothers. Before they set out, someone had the idea of making a guide rope for the boys. Several of the girls had worn aprons to school that morning and Miss Hunt collected these and tore them into strips. When all the apron strips were tied together they had a good length of rope. Fred would take one end and Miss Hunt would hold tight to the other. Then, when they found the farmhouse, the boys could follow the rope back to the pile and get the others.

Or, if they failed to find to find the Hinner place, at least they wouldn't get lost in the storm. One way or the other, they'd follow the apron rope back to the pile.

Fred and the two other boys went back out in the storm and began to walk around the straw pile in ever larger concentric circles. One time around, then a few steps back, and they'd circle it again. They shouted at the top of their lungs. They held out their arms hoping to brush against a piece of farm equipment or the side of a building. They stopped and peered through the vibrating air. Nothing. Not a sound aside from the keening of the wind. Not a solid shape anywhere. At least they had the rope of shredded aprons to guide them back to the straw pile. Without that the three boys would surely have drifted with the storm and frozen to death.

✦ ✦

At first May Hunt refused to despair. Now that the boys were back, they had eight voices. If they shouted in chorus, one of the Hinners would certainly hear them. They'd have to keep shouting to guide their rescuer through the storm. They kept it up as long as their voices held.

Nothing.

The smaller children had begun to whimper. Miss Hunt wondered if they'd be warmer if they could burrow farther into the straw pile. She asked Fred and his brother to get the pitchfork and dig them a deeper cave. Deep enough so that all of them could lie down together side by side. This would be their shelter for the night. Fred did the best he could with the straw. When May judged the cave large enough, they settled in. Without being asked or told, Fred took his place at the mouth of the cave—the coldest and most exposed spot. They all understood that it was he, if anyone, who would see them through.

None of the children had had anything to eat since noon. None of them had adequate clothing. Not a blanket. Few had even worn hats.

For a while they told stories and sang songs. Then Miss Hunt had the idea of calling the roll every few minutes. Anything to keep them awake. She believed—everyone on the prairie believed—that sleep was death. Fred Weeks? Present. Charles Weeks? Present. Mary Huntley? Present. Ernest Huntley? Present. Mabelle Huntley? Present. Frank Knieriem? Present. Addie Knieriem? Present. It passed the time. And it reassured her that all seven of the children were still alive.

❧ ❧

Sergeant Glenn estimated that there were twenty thousand people "overtaken and bewildered by the storm." Many of them were children sent home from school or out doing farm chores, but there were also farmers working in their fields or leading their cattle to water, doctors making their rounds, peddlers, salesmen, mail carriers, itinerant grocers. Glenn himself got lost as he tried to make his way home from the Huron weather station that night. Blowing snow sealed both his eyes nearly shut and he wandered around dazed and exhausted until by chance he encountered someone who pointed him in the right direction. There are hundreds of firsthand accounts of the onset of the blizzard, the wind shift, the first wave of blinding snow as fine as dust. There are scores of stories of narrow escapes, houses or barns found by accident or luck, horses or dogs that led their owners back to the barn, landmarks that suddenly appeared when the wind dropped for a moment.

Rare are the stories of rescues. People saved themselves or they weren't saved. With few exceptions, once a body was prostrate in the snow it stayed there.

❧ ❧

Had it not been for his eighteen-year-old half brother Will, that would have been Walter Allen's fate.

The afternoon was well advanced by the time the drays full of Groton's school children appeared on the main street of town. The teams came one behind the other, just as they had set out, only now the drays behind them were solid with huddled gray shapes. The horses labored, their noses clogged with melted and refrozen snow, their breathing rough. They moved up Main Street so slowly that the children had no trouble dropping off as the drays came abreast their houses. One by one, the kids landed on their buried feet, got their balance, and then waded, staggering through the drifts and bent nearly double in the wind, into the arms of relieved parents. Slow as they were, the horses didn't take long to cover the few blocks of town. And of course the loads got lighter as the kids peeled off.

Every child was safe now. For the men who had volunteered to rescue the kids there was still the business of getting the horses unhitched, awkward to do with frozen hands, stowing the drays so they didn't get buried under snowdrifts, and putting the horses under shelter for the night. A nasty job in a storm like this, but it wouldn't be any easier tomorrow.

The men were in the middle of the chore when W. C. Allen and his boy Will came hustling up. Shouting over the wind, Allen told them there was no sign of Walter. Had the teachers taken a head count before they set out? Maybe the boy had stayed behind in the schoolhouse for some reason. Or could he have fallen off one of the drays on the way home? Edna was worried sick. They'd have to go back—and go fast while there was still light.

The men conferred for a few minutes and decided they better take two drays for an extra measure of safety. W.C.'s boy Will would come along, too. At eighteen he was full-grown and as strong as any of them.

If the men who had driven the drays thought it was a fool's errand to go back out in a blizzard like this, they kept the thought to themselves. Each knew, or felt, that there really wasn't any choice. There was a boy missing. They'd never be able to look W.C. or Edna Allen in the eye if they didn't at least try to look for their son.

It was harder going back because it was darker and they had to drive the horses into the wind. The men kept their heads down and their shoulders hunched up almost to their ears. No need to talk even if talking had been possible over the wind. Will was the only one among them who wasn't grim faced and hollow chested. He was looking forward to giving his brother a good tongue-lashing. He'd be teasing Walter about this for years to come: "Remember the time that damn fool of a boy ran off and hid in a blizzard and scared us all half to death? Just like Walter to pick the worst storm in history. The terrible blizzard of '88."

They pulled the horses up next to the main schoolhouse door on the east wall, the lee side. The tall, narrow building was taking quite a beating in the wind, but it wasn't so bad in the steep-sided hollow that the wind was scouring out behind the school. They stood there for a minute watching the snow whistle off the lip of the drifts. Not a track in sight, not a sign of the boy. They went into the school and hollered and tramped through the classes just to make sure. The cold was already bone-chilling in the empty rooms.

The men went back outside and spread out around the school. There was nothing, no one. It was getting more painful every minute to keep their eyes open. After a while they stopped looking and stood by the school entrance shouting over the wind. W.C. said he was going to stay and keep looking. The others should take the drays back. He'd find the boy and get him home.

Or die trying, they all thought but nobody had to say.

The other men wouldn't hear of it. W.C. would go home with them. He had a wife and family to provide for. It was madness to stay out in the storm alone.

There was more shouting. Finally W.C. agreed, or at least stopped arguing.

The men were so intent on their shouting match that they'd forgotten about Will Allen. No one saw him disappear. Strangely, no one missed him on the way back. W.C. was sunk in misery and numb with the cold. The other men just wanted to get out of the storm before it was too late. Their hands were like blocks of ice. It was impossible to hold the reins. They let the horses go and hoped for the best.

Arthur E. Towne was a boy when his family took him from Vermont to Huron in Dakota Territory during the summer of 1881. That first winter a blizzard struck Dakota, a bad storm that left the prairie for miles around littered with the corpses of cattle and horses. Arthur's father had the misfortune of getting caught outdoors in that blizzard, but being a Vermonter he was canny about snowstorms. When he couldn't see which way to go, he dropped to his hands and knees and crawled along the ground, feeling for the wagon track as he went. "He had been nearer death in those few minutes than he ever was again while he lived in the west," his son wrote. His fingers were in agony by the time he made his way back to their sod house, but Towne survived.

That was how Will Allen found his brother Walter. Somehow Will figured out that the visibility was a little better right next to the ground—"under the storm," as the settlers said. So he fell to his hands and began to stalk around like a dog. Whether it was instinct or pure dumb luck, Will eventually stumbled upon Walter's body.

The boy was unconscious but breathing. His limbs hung limp from his frame, which was good because it meant they weren't frozen.

Will barely felt the cold as he carried and dragged his eight-year-old brother back to Groton. Instinct or luck stayed with him. If he lost his way they'd both be dead. If he dropped the boy, or tried to leave him and go for help, it was unlikely that he would ever find him again. Nightfall was near.

The Allens' house was at the upper end of Main Street, and next door to the house was the building that housed W.C.'s law office. That was as far as Will got. Though he knew his parents would be waiting at home in panic, he couldn't summon the energy to cross those last few steps. Will dragged the boy inside their father's office, out of the wind and cold, and then his strength gave out.

W.C. and Edna were beside themselves when Will appeared at the door. In a minute he told them what had happened. Walter was still next door in the office, conscious but too weak to walk. Will was barely able to stagger over there again and help his father get the boy back home.

Walter had his eyes open and kept looking from his father to his brother and back, but he still could not understand how he had gotten from the snow-packed ground outside the school to his father's office. All he knew was that he had a terrible stinging sensation all over his skin and a feeling of cold that penetrated to the core of his being. He was shivering uncontrollably.

Walter was too weak and dazed to be aware of a little trickle running down the side of his leg. It was the water in the perfume bottle that had frozen while he was lying unconscious in the snow. Frozen and burst. In the warmth of his father's office, the ice had thawed and the water was seeping through the shards of the glass.

Grace Rollag was stunned to see the familiar bulky shape of her husband Ole heave itself out of the storm just as she and Peter returned home from the barn. At first she didn't trust her eyes. With the snow blowing sideways like this and her eyes crusted in ice she couldn't be certain. Shapes kept emerging and then the snow snatched them back again. And the sounds that flew by in the torrent—mutterings and keenings and distant shouts whose words she could almost make out but not quite. But yes, it was Ole sure enough. Snow was so thickly matted in his beard and eyebrows that she could barely see his face. Where had he come from? Where was Charley? *"Bare fint*—fine, fine," he shouted to her over the wind. Once they were inside and Ole got the ice off his beard, he explained how he had left the boy with the neighbors because his face had been so badly frozen. He'd be fine there overnight. The horse, too.

"It was a miracle that Ole had found the way," Grace wrote later. "But he never lost his composure so that the whole time he knew which direction to take. And we were happy indeed when at last we were all inside the house."

❧ ❧

By nightfall an old Norwegian settler named Ole Tisland had rubbed the ice around his eyes so often and so violently with the back of his sleeve that he had torn the frozen skin away and exposed his cheekbones. A father and grandfather who had come to America after the Civil War, Ole had been out shoveling on his farm in Brookings County in southeastern Dakota when the blizzard struck. His son-in-law, Bengt, saw Ole standing there with his shovel in hand staring fixedly at the northwest sky. Ole's wife, Karen, saw him, too, and wondered why he didn't rush for home.

That night a little boy named Carl Hildens, who lived on the farm adjoining the Tisland place, heard someone shouting outside. Carl and his parents were sitting by the stove while the blizzard raged on. The boy told his parents he heard shouts for help—he insisted over and over that he wasn't imagining it. Finally Carl's father, Hans, got his overcoat and went outside. He called at the top of his voice, stopped to listen, called again. Nothing. Nothing but the scream of the wind and the snow lashing his face. "God help anyone lost in this," Hans told his son. "It is getting colder."

❦ ❦

There is no way of knowing how many people spent the night out on the prairie. Or when the first person died.

The dark was absolute by six o'clock.

Hundreds, possibly thousands of children spent the night in schoolhouses or the boarding places of their teachers while their parents sat up and wondered where they were. Precious desks and tables and chairs were fed into the stoves. The last bits of lunches were carefully divided. Thin coats served as blankets.

These were the lucky children.

Among the unlucky children that night were Lena Woebbecke, lying on the side of a hill near Seward, Nebraska; the five Schweizer boys from school number 66 in Dakota's Rosefield Township—Johann, Heinrich, and Elias Kaufmann, Johann Albrecht, Peter Graber; the seven students of May Hunt huddling in a haystack near the Hinner farm; the two Westphalen sisters, Eda, thirteen, and Matilda, eight, who had left their country school near Scribner, Nebraska, together and lost their way in the storm; the Heins children, who had unlocked their schoolhouse door and run off into the storm; Etta Shattuck in her haystack in Holt County, Nebraska; sixteen-year-old Omar Gibson and his twelve-year-old niece, Amelia Shirk, pressed together under a horse blanket in the south part of Holt County; nine-year-old Roman Hytrek, who had lost his coat on the open prairie as he wandered alone with his dog; seven-year-old John Schaaf, who had been sent off on a pony to do an errand for his parents and was overtaken by the storm.

As night fell, each one of them had to understand that there was no hope left of finding shelter, no chance of being found. The unimaginable had happened.

CHAPTER SEVEN

God's Burning Finger

It was the turbulence kicked up by the strengthening low that bewildered those caught out on the prairie that afternoon. But it was the high that dropped down in the wake of the low that killed them.

By the twelfth of January, the arctic air that had been massing for days under the persistently clear skies of western Canada had matured until it resembled a vast shallow rotunda many hundreds of miles across but less than a mile and a half deep at the center—a dome of high pressure, in meteorological parlance, though what this familiar phrase fails to capture is the dynamic fluid nature of a high-pressure system. The cold dense air that raised barometric pressure at the surface not only sank slowly, but also spread out horizontally in currents radiating from the center of the high. The effects of the earth's rotation turned these air currents to the right in a vast clockwise outward spiral known in meteorological jargon as an anticyclone. Just the opposite happened in the low, or cy-

clone, that preceded the high: Winds spiraled inward toward the
center of lowest pressure in a counterclockwise pattern.

As the anticyclone surged south down the center of the continent, the dense frigid air encountered the great spine of the Rocky
Mountains rising along its western edge. Clockwise circulation
pushed the southwestern sector of the air mass up against the eastern flank of the mountains, and the mountains pushed back. The
pressure kept increasing. All through the twelfth, the cold high
continued to build into Alberta and northern Montana even as Signal Corps observers in Dakota and Nebraska tracked the progress
of the advancing low on their plunging barometers. Though hundreds of miles separated them, the low and the high were now
locked together and violently reinforcing each other. By midafter-
noon the pressure difference between the center of the low over
southeast Nebraska and the center of the high over Edmonton in
Alberta was 1.2 inches of mercury—a pressure gradient rarely seen
in the course of the most severe winter.

As night fell, the low was still deepening, and as it deepened it
sucked in greater and greater volumes of air. Some of that volume
got fed into the high. Picture it like shoveling snow into an ever-
rising mound at the center of a gargantuan circular drive, but instead of snow, the atmosphere was shoveling the air itself. The low
was shoveling in air at the lower levels of the atmosphere, sucking
it up in an ascending spiral, and spewing it out aloft. Some of what
got spewed out of the low in the upper atmosphere landed on the
high, effectively piling and cramming even more air onto the top of
the dome. In meteorological shorthand: Upper-level divergence
over the developing low worked in tandem with upper-level con-
vergence over the strengthening high.

The lower the air pressure fell over Nebraska and Iowa on the
night of the twelfth, the higher it rose over Montana and northern
Dakota. The greater the pressure difference between the cyclone
and the anticyclone, the stronger the winds blew in a futile effort to

fill the vacuum created by the low. Northerly gales dislodged some of the coldest air on record from central Canada and forced it south over the Upper Midwest. As long as the high kept building in behind the low, the arctic air would keep moving south, steered and funneled by the Rockies, right down the dead-level center of the country, until it crossed the border and spilled into Central America.

Only the most intense cold waves got that far south. As January 12 drew to a close, it was becoming clear that this was shaping up to be one of the most intense cold waves ever seen in the United States.

※ ※

A sign of the fierceness—and strangeness—of this storm was the eerie electricity that crackled through the air as the temperature began to drop. It was like a lightning storm, only instead of bolts flashing thousands of feet between cloud and ground or cloud and cloud, smaller electrical discharges sparked at the surface. In effect, the storm had created a kind of horizontal thunderstorm—with one key difference: While a thunderstorm is driven by the instability of rapidly warmed air, the driving force behind these static discharges was the extreme differences between the various currents of air that were screaming past each other in the blizzard.

As the winds roared from the high to the low, air saturated with a mixture of shattered snow crystals, ice pellets, and water droplets blew over surfaces that were much colder or warmer than the air itself—crusted snow on the ground, patches of bare earth or vegetation that had absorbed the warmth of the morning sun, the sides of frame or sod houses. The storm was developing so rapidly that seams and pockets of sharply contrasting temperatures were rubbing up against each other. Think of the cold front sweeping down not like a smooth plate-glass window, but rather like a curtain crimped into deep irregular folds, with very different kinds of air caught up between and within the folds. To add to the volatility, the cold air immediately behind the curtain was quite shallow: It spread over the prairie like spilled syrup oozing across a countertop, a thin film of cold sliding under the much warmer air above and around it. Cold and warm, liquid and frozen, humid and dry were abruptly dragged together by the developing storm. Enhancing the roiling contrasts were the airborne snow and ice particles themselves, which were of markedly different temperatures and structures, warmer or colder, wetter or drier, depending on which part of the storm they came from. All of this acting together set up the ideal conditions for significant static discharge at the surface.

As the patchwork curtain ripped over the prairie, the very air began to tingle with electricity. Sailors used to call this phenomenon Saint Elmo's fire, or the corposants (holy bodies), and meteorologists have since renamed it corona or point discharge. Very likely, the electrical field at the surface was enhanced by lightning aloft—"thunder snow" is another sign of a violently intense cold front. But since the visibility was so poor and the wind so loud, no one saw the lightning bolts or heard the booms of explosively expanding air.

The Saint Elmo's fire, however, was unmistakable and terrifying. At the height of the blizzard people inside their houses felt their hair rise off their scalps. They watched showers of sparks leaping off the ends of pokers as they held them to their stoves. The air popped and sizzled when a hand was passed over someone's head. "Frank went to put up the draft in the stove," wrote Dakota pioneer Julia Carpenter of her husband, "when his fingers snapped and fire came from them. He tried this again and again, putting first one finger and then another near the stove, always with the same effect. I did not try it. When Frank was out of the room there was still a snapping."

At Fort Buford near Bismarck, Signal Corps observer Sergeant A. Schneider wrote that the air was charged with electrical sparks "more than an inch long" that jumped from the telegraph wires to the switchboard. As Schneider moved his hand toward the switchboard, a spark jumped off the ends of his fingers when they were about two inches from the board and nearly knocked him to the ground. Panicked, he tried to leave the station, only to find that heavy drifts had blockaded the door. Schneider finally escaped through the window.

Sergeant Glenn in Huron spent the evening at the weather station watching sparks of electricity leap from the gilt molding used for hanging pictures and play over the gilt stamping of the curtains, the stove, and the chains from which the lamp fixtures were suspended. Glenn was literally stunned when an electrical shock passed from his hand to the hand of an acquaintance standing just outside the office. Many settlers became so terrified of the sparks showering off their stoves that they would not go near them, even to add more fuel.

Herman Melville described Saint Elmo's fire in "The Candles" chapter of *Moby-Dick* as *The Pequod* labored through a typhoon east of Japan: "All the yard-arms were tipped with a pallid fire; and touched at each tri-pointed lightning-rod-end with three tapering white flames, each of the three tall masts was silently burning in that sulphurous air, like three gigantic wax tapers before an altar." To the hapless mariners it appeared as if "God's burning finger has been laid on the ship."

It was after God withdrew His finger that the film of cold air pooled and deepened and the temperatures over the prairie began to plunge in earnest.

The storm was just grazing Saint Paul with light dry snow and bursts of wind when Lieutenant Woodruff returned to the Chamber of Commerce building from his evening meal at 10:15 P.M. Central time. The temperature in Saint Paul had been rising all day, and by the nighttime observations it had reached 11 above zero. But Woodruff could see at a glance on the afternoon maps that the cold was coming his way fast. Indeed, when the telegrams of the 10 P.M. (Eastern time) observations began to arrive, it was clear that the cold was on his doorstep. Fort Assinniboine was now reporting 27 below zero with north winds at 13 miles an hour; 4 below in North Platte; Omaha, which had reached 27 above at 3 P.M. Eastern time, was reporting temperature of 6 below with winds of 30 miles an hour. Eleven below in Yankton. Seventeen below in Huron. The cold had dropped all the way south to Kansas, where Leavenworth was reporting 4 above, a drop of 29 degrees in the past seven hours, with northwest winds at 22 miles an hour.

Even in a region notorious for sudden cold outbreaks, these numbers were impressive. In the course of four and a half hours the temperature fell 50 degrees at Helena. At North Platte, Nebraska, a drop of 32 degrees in thirteen hours. At Keokuk, Iowa, 55 degrees in eight hours.

"The most decided and most severe cold waves follow severe storms," Woodruff had written in "Cold Waves and Their Progress." "The maximum effect, or the minimum temperature, occurs when the barometer is the highest above normal." By midnight on the twelfth, it was clear to Woodruff that this Dakota storm was bringing down one of the most decided and most severe cold waves he had ever seen or was likely to see. What path it would take, how far south the cold air would spill, whether it would engulf the cities of the East Coast—these were questions Woodruff knew he was unable to answer with any degree of confidence. Luckily, he didn't have to. The cotton- and sugar-growing regions of Louisiana and Texas, which were particularly sensitive to sudden cold waves, and the hubs of Chicago, Boston, New York, Philadelphia, and Washington were Greely's headache.

Shortly before midnight, Woodruff issued his indications for the following day—unlucky Friday the thirteenth, as it happened:

January 13, 1888 12:15 AM
Signal Office War Department Saint Paul
Indications for 24 hours commencing at 7 AM today

For Saint Paul and Minneapolis: Snow colder with a cold wave, fresh northerly winds

For Minnesota: Colder with a cold wave, snow followed in northern part by fair weather, fresh northerly winds

For Dakota: Local snows, colder with a cold wave, fresh northerly winds becoming variable.

In a few hours it would become clear how inadequate the word *colder* was to describe the air that was pouring down over the prairie on those "fresh northerly winds."

CHAPTER EIGHT

Exposure

The fear came first, but the cold followed so hard on its heels that it was impossible to tell the difference. The two smaller boys, Heinrich and Elias Kaufmann, felt it first. They were *lost*. Their minds might try to shove the word aside, but their bodies knew it was true. Mr. Cotton and Peter Graber's two younger brothers were gone. There was no sign of the Grabers' farmhouse at the end of the field. No sign of a barn or a haystack or a fence in the blinding snow. They were lost, the five of them together. They could see from the way the older boys—their brother, Johann, and Peter Graber and Johann Albrecht—stopped and cupped their hands around their eyes and turned in circles that they had no idea where they were, either. Elias, who was only seven, wanted to scream for his mother, but he knew that the others would be angry at him. And what would be the use? Hadn't they all been shouting at the tops of their voices?

In the first flush of panic Elias felt his stomach turn to water and

the cold sweat drip down his ribs and pool at the small of his back.
He had worn no hat or gloves or heavy woolen coat when he left for
school with Heinrich and Johann in the mildness of the morning.
None of them had. The wind found every gap in the homespun
cloth, every pore in the woolen socks and underwear, every buttonhole and cuff. Elias hunched his bony little-boy shoulders. Pinpricks
of snow lanced his clammy skin and made it tingle. The blood-rich
capillaries began to tighten down at his feet and hands and his exposed neck: His body was trying to conserve warmth by removing
blood from the lash of the wind and sending it deeper within.

Lost, alone with his terror, too scared to cry, Elias Kaufmann
started to shiver.

In an awful way, the five Schweizer boys who had wandered off
when the blizzard struck had become factors, very small and frail
factors, in the immense equation of the weather. Physics dictated
that their warm body tissues and fluids would eventually reach
equilibrium with the cold fluid of the ambient air. It was biology
that infinitely complicated the equation. The size and shape,
weight, consistency, and age of their bodies; their gender; the actions, both voluntary and involuntary, that their brains initiated;
the subtle physical and chemical changes tripped by their emotions—all of these would determine how long they survived. Every
living thing fights the physics of freezing to death, whether it
wants to or not. Whether the body wins or loses is so complicated,
so mysterious a process, as to resemble fate or luck.

When they left the schoolhouse and walked out into the storm,
the internal temperature of the boys was presumably 98.6 or close
to it. They may have gained a couple of degrees in the first minutes
outdoors from the force of their exertions to remain upright in the
wind and from the adrenaline rush of striding off into the streaming snow ahead of their teacher. But inevitably, gradually at first

and then more swiftly, their core temperatures began to drop as their bodies lost heat to the air. The same basic physical processes that move heat energy through the atmosphere—radiation, conduction, evaporation, and convection—worked against the boys.

Nearly half of their bodies' total heat production radiated through their uncovered heads in the form of infrared energy— invisible wavelengths of electromagnetic energy that travel through the air until they are absorbed by an object in their path. Radiational cooling, as the process is called, is what created the frigid air mass of the cold wave in the first place: During the run of calm, clear nights up in Canada earlier in the month, temperatures dropped quickly at the surface of the earth as the long waves of infrared energy radiated into the atmosphere. With no clouds to reflect the long waves back, they just kept rising and radiating heat up and out into space. On a much smaller scale, the same thing was happening to the body heat rising off the boys' heads.

Their wet clothing sapped away additional body warmth through evaporation and conduction. In order to evaporate—in other words, to change state from liquid to gas—water must absorb heat. Sweat cools the body by whisking away warmth as it evaporates off the skin. The water in the boys' wet clothing acted the same way as sweat does, and the wind greatly exacerbated the heat lost to evaporation. What made it worse was that their clothing never dried because the wind kept plastering them with snow, which the warmth of their skin melted. The wet absorbent material of their shirts and trousers and underwear soaked up even more heat.

Conduction, the process that moves heat directly from one object to another, wicked heat from the surface of the boys' skin to the cold water that saturated their clothing. Since heat is conducted twenty-five times faster in water than in air, the most mortally efficient conduction occurs when a person falls into cold water. Even water at 60 degrees is cold enough to quickly induce a condition known as immersion hypothermia. The boys carried their own frigid streams of water around their bodies. Later, when they knelt to pray, more heat would be conducted from their legs into the snow.

Most insidious was convection, the process that carried off the small envelope of heated air next to their skin. Convection, the transfer of heat by the movement of the air itself, accounts for why hot soup cools faster if we blow on it: Our breath carries off the hot, steamy air rising off the surface of the soup and replaces it with cooler air; as the soup transfers its heat into the flow of our breath, it cools. In the blizzard, the incessant wind was the breath that stripped away the warmth of the boys' bodies and mixed it into the colder air around them. For a time their bodies kept producing more heat, and the wind kept blowing it away and replacing it with cold air. But eventually they started losing heat faster than their bodies could make it. Again, there is an analogy with water: An icy current would have acted in the same way had the boys fallen into a river. In a sense they were being swept along in a river of air—and the swifter the current, the faster the thin shells of their body heat were peeled from them.

An American explorer and scientist named Paul Siple worked out a formula for how convection kills through a combination of cold and kinetics. He called it the windchill factor. Cold, of course, can kill without wind, but it takes longer. Wind and cold together accelerate the process. To be precise, the amount of body heat lost to wind increases as the square of the wind's velocity. Air moving at 8 miles an hour saps the body of four times more heat than air moving at 4 miles an hour; at winds of 20 miles an hour, twenty-five times more body heat is lost than at 4 miles an hour. Paul Siple had a lot of experience with wind and cold because he spent years working as the biologist on the Antarctic expeditions of Admiral Richard E. Byrd during the 1930s and '40s. It was in the Antarctic that he calibrated windchill through a series of experiments with water in plastic cylinders. Siple's windchill index has since been revised, updated, and recalibrated, most recently by the National Weather Service in 2001, but the fundamentals remain.

Television weathercasters like to say that windchill is what the weather *feels* like. Using the 2001 windchill index, when the wind is blowing 30 miles an hour at a temperature of 25 degrees, it *feels* like 8 degrees. "Feels like" is a fuzzy term for an exact transaction. What windchill means is that it's irrelevant that the thermometer reads 25 degrees: If the wind is blowing at 30 miles an hour, the exposed parts of your body are losing heat at the rate that they would if the temperature were in fact 8 degrees.

When the Schweizer boys left school late in the morning, the windchill was about 5 degrees above zero. At 9 P.M., four hours after the sun set, the windchill had dropped to 40 below zero. In conditions like that, exposed human flesh freezes in ten minutes.

Ten minutes to turn warm skin and blood to ice. The five boys had been outdoors by that point for over nine hours.

※ ※

For a while, shivering kept them warm. The twitching started in their faces and necks and moved down their torsos and out to their arms and legs, rhythmic waves that rippled through their muscles at a frequency of six to twelve cycles every second. As long as their flesh jumped and danced around their bones, their muscles were producing enough heat to keep their bodies warm. Inside, their vital organs went on working normally. But shivering cost them dearly. As they shivered, their consumption of oxygen doubled or tripled, a sign of accelerating metabolism. Heinrich and Elias quickly burned through the calories of their last meal—the coarse bread sweetened by syrup or jam they had carried with them to school that morning. When that was gone their bodies looked for more fuel with which to combat the cold. But there wasn't any—at least nothing they could convert to heat quickly. The boys had expected to be home by now, with a fire and a smiling mother to cook their supper, so they had brought no food but their meager lunches.

Shivering on an empty stomach is like burning your clothes in the stove once the coal and furniture are ash. The energy it took to walk, just to remain upright in the wind, made their body heat dissipate even faster. Fear threw open still more vents. Even mild mental stress hastens heat loss. Terror and exhaustion are as efficient as wind in scouring heat out of the human body.

The Schweizer pioneers frowned on complaining. Their children were raised to be cooperative, to think of others before themselves, to work together for the good of the family and the group—above all, to be humble before God. So it came as a shock to the older boys when Heinrich and Elias started falling behind and whining. They were hungry. They couldn't feel their feet or hands. Their eyes stung from the blowing snow.

Johann heard his younger brothers wail over the shriek of the wind and could hardly believe his ears. Crying like babies, screaming for the others to wait for them. Johann saw Elias fall as he stomped through the drifts trying to catch up. Or perhaps Heinrich had pushed him. Then Heinrich stumbled. Both of them were angry and red-eyed. The bickering and crying were signs that the younger boys' bodies were beginning to succumb to the cold.

Irritated as he was, Johann would never have dreamed of abandoning his brothers. The five of them must stay together no matter what. Johann motioned to Peter Graber and Johann Albrecht to stop. The three older boys put their faces together so they could hear each other and talked about what to do. Peter Graber and Johann Kaufmann, the two sixteen-year-olds, took charge. They knew it was essential to keep moving and continue looking for shelter. If the little ones couldn't go on they would carry them. Johann and Peter bent over at the waist and had the small boys climb onto their backs. Now there were three sets of tracks instead of five.

For a time, the exertion did Johann and Peter some good. Or seemed to. The effort of carrying the extra weight sent a ripple of warmth through their bodies. And shivering contributed its own pittance of heat—for despite the exertion, it was so cold in the wind and their clothes were so thin that they still shivered. Shivering and labor combined to work and warm their muscles. But it was warmth they couldn't afford for very long. The boys were paying for every step and every spasm—paying with precious currency. Without food, without relief from the wind and cold, they were soon bankrupted.

Whenever they stopped moving even for a moment, to catch their breath or peer into the impenetrable air, the tide of warm blood ebbed from their skin and extremities and flowed inward to the centers of their bodies. But the tide was no longer warm enough to boost the temperature inside them. The heat of exertion rose off them and dissipated like steam. Gone. Their core temperatures, the temperature of the blood near their hearts, began to drop. At 95 degrees they exhibited the first signs of mild hypothermia.

Strangely, their minds were affected before their bodies. The boys became peevish. They wanted to argue, but when they opened their mouths to shout over the wind, they had trouble forming the words. Thoughts came slowly and only with exaggerated effort, like moving under water. For the first time, they blamed each other for wandering off from Mr. Cotton and the two younger Graber boys. The three who were still walking stumbled, and they saw each other stumble, and for some reason this filled the boys with annoyance and disgust. Shivering reached a climax as their bodies clung to an internal temperature of 95 degrees, but the uncontrollable twitching of their flesh disgusted them as well.

It is during this first mild stage of hypothermia that mountaineering and exploring parties start to bicker and group solidarity breaks down. Ordinarily docile and cooperative individuals turn waspish and vindictive; leaders make bad choices. Everybody thinks someone else is to blame for the misery of being out in the cold. Mountaineers call it "cold stupid." The dulled mind begins to throb around a single image—really more a sensation than an image: the craving for warmth.

Johann and Peter were now shivering so violently that it was difficult to hold Heinrich and Elias on their backs. What made it worse was that the younger boys had gone limp on them, like half-empty sacks of peas. Johann kept shouting for his younger brother to grab tightly to his shoulders, but the little boy made not the slightest effort. Again and again he slid to the ground. He just sat in the snow staring dazedly when his brother bent over to pick him up again. Johann could hardly restrain himself from smacking the child.

What Johann did not realize was that Heinrich and Elias had become dangerously chilled while he and Peter were carrying them. Lying inert against the older boys' backs, they had fallen into the apathy of deeper hypothermia.

At outside temperatures of 35 below zero, the body loses a degree of heat every thirty to forty minutes—and far more rapidly than that if the clothing is wet. By evening the windchill temperature began to approach that level in Dakota. As their core temperatures dropped degree by degree, the boys' minds betrayed them more severely and bizarrely. With their body temperatures at 93, amnesia began to cloud their thoughts and impair their judgment. Thoughts oozed slowly out of their brains, and time itself seemed to drag on leaden feet. This is the temperature at which people make foolish, sometimes fatally foolish, decisions. They take the wrong path. They refuse to turn back from their attempt to reach the mountaintop. They lose gloves or hats or discard precious supplies. It's like being insanely drunk with cold. From here on, the boys would remember nothing.

When their internal temperatures hit 91 degrees, they ceased to care what happened to them. Their speech became slurred. Johann Kaufmann and Peter Graber and Johann Albrecht were now as dull and apathetic as the limp younger boys cradled in their arms. Their bodies were so cold that nerve impulses moved sluggishly to muscles, and the muscles failed to respond normally once the impulses reached them. The boys stared down at the frozen blocks of their hands and wondered dimly why they couldn't make them move or feel anything. They had trouble contracting the muscles in their thighs and calves that they needed in order to walk; but once the muscles contracted, they couldn't relax them again. The boys stumbled and staggered. It was the wind that determined where they went, not their numbed brains.

As darkness fell, Maria and Johann Albrecht took some small comfort in the fact that the teacher, Mr. Cotton, hadn't come back either. They had gone over it again and again through the long day of waiting and wondering. Mr. Cotton, who was boarding with them, must have stayed at the schoolhouse with their son Johann and the other children. That would explain why neither he nor the boy had come home. Albrecht reassured his wife that this was the only possibility. He had been outside himself when the storm hit and he knew how bad it was. Albrecht had never seen anything like this in all their years in Dakota. It would have been madness for the teacher to dismiss school and send the children out into this storm.

Maria Albrecht tried to believe her husband. She agreed that they must not even think of traveling to the school to get Johann until the wind had died down. Now in the darkness the storm was more dangerous than ever. Together they would pray and wait for the morning. But still, whenever she saw Johann's younger brother Peter, Maria could not stop herself from sighing and shaking her head. "O where is my child?" she wailed. "My heart is going to break."

Shivering is a very demanding way of warming the body. But the body shivers as long as it's able to because the alternative is much worse. Shivering is the body's last defense against the abyss of deep, potentially fatal hypothermia. Once shivering stops, the chilled body falls quiet and then shuts down rapidly. On a graph of temperature loss plotted against time, the drop from 98.6 to 90 looks like an intermediate ski slope; below 90 is a cliff.

Heinrich and Elias stopped shivering first; then Johann Albrecht; finally Johann Kaufmann and Peter Graber. Their core temperatures were now around 88 degrees. Severe hypothermia had set in.

The functions of their vital organs slowed. The chilled blood thickened. Their hearts turned stiff and frail as the cold penetrated deeper. Like the muscles of their legs and arms, their heart muscles failed to respond efficiently to nerve impulses. Contractions became weaker and weaker; the pumping action was barely forceful enough to push the viscous blood through their veins. A vicious cycle set up as their weakened hearts failed to supply the tissues with the oxygen they were craving: The lack of oxygen made their bodies unable to complete the metabolic cycle, causing lactic and pyruvic acids to accumulate in their tissues; the buildup of these acids made their hearts beat even more feebly and erratically, which in turn spiked levels of the acids.

The deepening cold radically redistributed their bodily fluids. During the first hours, the blood had retreated from their skin and extremities into the core of their bodies to keep the central organs warm. The temporary rise in the volume of blood flowing through vessels deep inside increased the production of urine. Soon the boys were desperate to urinate, but their hands were so paralyzed by cold that they couldn't open their flies. Eventually, as their bladders emptied repeatedly and their core temperatures kept falling, their blood volume began to decrease. The blood itself became increasingly viscous as more and more water was retained in their tissues. Their kidneys were no longer able to conserve water. The boys urinated again and again, probably wetting themselves and adding to their misery. Dehydration became acute, and this in turn made their blood volume sink even lower.

There was a measure of protection in this shutting down of their internal processes, at least for a time. As their metabolism slowed, their brains required less oxygen, which was fortunate since their bodies were incapable of supplying much. Doctors today routinely induce extreme hypothermia during certain open-heart operations by pumping very cold blood into the body with a heart-lung machine. For an hour or so, the doctors keep the body at the threshold of death while they operate on the cold, motionless heart. The stilled pulse and the barely functioning brain reduce the patients' risk of heart attack or brain damage. Had the boys been rescued at this point and warmed properly, they might have recovered fully from an internal temperature of 88. It's dangerous to be this cold but not necessarily fatal.

But the odds of being rescued diminished steadily as the day wore on. The Kaufmanns and the Albrechts still believed that their sons had stayed in school, as they had been told to do again and again. And the Graber parents, who knew the truth, were too frightened to go out searching in the storm.

At 87 degrees, the boys probably would not have recognized their parents anyway.

Below 87 degrees, they began to lose their grip on reality. They ceased to know or care that they were cold. They gave up looking for shelter.

Hallucinations and delusions set in. Starved of oxygen, unhinged by stress and fatigue, the brain fabricates its own reality—often the fulfillment of desperate prayers. Two hikers stranded on Everest at 28,600 feet in temperatures of −35C hallucinated together that their supply officer had joined them in their bivouac—the only person in the team who might have brought them dry sleeping bags, food, and oxygen. Sailors who survived the sinking of the USS *Indianapolis* in the eastern Pacific in July 1945 hallucinated that an island was within reach. Some of the men stripped off their life jackets and drowned. The Hans Christian Andersen fairy tale "The Little Match Girl" is a classic case of wish fulfillment hallucinations induced by severe hypothermia: a small girl, shivering barefoot in her snow-dampened rags, watches in amazement as a series of fantastic images appears out of thin air every time she strikes a match—a warm stove, a table laden with food, a Christmas tree glittering with candles, the return from the grave of her recently deceased grandmother. The next morning, passersby find the girl dead, with a smile on her face.

Johann, Heinrich, and Elias Kaufmann, Peter Graber, and Johann Albrecht, brought up together in the tight-knit, deeply religious world of the Ukrainian Mennonites, may have hallucinated that Jesus came down in his flowing robes to take them to heaven. Or that their mothers walked out of the storm bearing plates of hot poppy seed cake or the savory dish they called *Käseknöpfle*—literally, cheese buttons—made of delicious dough and cheese and onions.

People freezing to death sometimes find they are unaccountably happy and relaxed. They feel flushed with a sudden glow of well-being. They love the world and everything in it. They want to sing. They hear heavenly music. As the mind and the body amicably part company, the freezing person looks down on himself as if he's hovering overhead or already in heaven or a returning ghost. There is his body, lying miserable in the snow, but somehow he is no longer trapped in it. He's gazing at his corpse and walking on. He's telling the story of his miraculous escape. In his account of the 1996 climbing disaster on Mount Everest, Jon Krakauer wrote that in the depths of exhaustion and hypothermia he experienced a "queer detachment" from his body, "as if I were observing my descent from a few feet overhead. I imagined I was dressed in a green cardigan and wingtips. And although the gale was generating a windchill in excess of seventy below zero Fahrenheit, I felt strangely, disturbingly warm."

This disturbing warmth is another common sensation in advanced hypothermia. Right before the end, the skin may feel like it's on fire. The bliss of merging with the cold is interrupted by a sensation of burning and suffocating. Doctors are not sure why this happens. It may be a delusion manufactured by the oxygen-starved brain or it may be that for some reason in the last minutes of consciousness the body sends a surge of blood back to the constricted capillaries at the surface of the skin. Whatever the cause, the result is that victims of hypothermia suddenly feel so hot and stifled that they strip off their clothes.

A few days after the blizzard, the *New York Tribune* reported in perplexity that a number of the dead were found with torn or missing clothing—collars ripped away from their throats, hats tossed off. The paper's reporter speculated that storm victims had suffocated on the fine powdery snow and wind-borne ice pellets: "In a genuine blizzard the air is filled with fine ice dust, driven with terrible force, which chokes the unfortunate victim in a short time if he attempts to stand against it." But the *Blue Valley Blade* in Nebraska correctly attributed this phenomenon to delusion and hallucination: "At this stage of freezing strange symptoms often appear: as the blood retires from the surface it congests in the heart and brain; then delirium comes on and with it a delusive sensation of smothering heat. The victim's last exertions are to throw off his clothes and remove all wrappings from his throat; often the corpse is found with neck completely bare and in an attitude indicating that his last struggles were for fresh air!"

It sounds bizarre—to wantonly sacrifice warmth and cover just when you need it most—but it's common enough that doctors have given the impulse a name: paradoxical undressing. Before paradoxical undressing was identified, police routinely mistook hypothermic women with torn or missing clothing for victims of sexual assault. The reaction explains a disturbing incident in military history. After a brutal three-day storm in January 1719, hundreds of Swedish soldiers were found stripped and dead in the field in the wake of a disastrous campaign against Norway. At the time it was assumed they had been plundered by their comrades, but now doctors believe that they tore off their own clothes as their minds and bodies went mad with cold—a mass outbreak of paradoxical undressing.

❧ ❧

For the five Schweizer boys, the end was probably peaceful. "First—Chill—then Stupor—then the letting go—" as Emily Dickinson imagined freezing to death in her poem "After great pain, a formal feeling comes." As their internal temperature dipped below 85 degrees, the hallucinations lost their grip, the imagined music stopped, the sensation of freezing or burning faded. They just wanted to go to sleep. "I was getting so terribly tired," wrote another boy who had gotten lost in this storm at the age of thirteen. "I felt sleepy. I thought if I could only lie down just for a few minutes I would be all right. But I had heard the farmers telling stories about lying down and never getting up again in snow storms. So I kept on, but I finally got to the point where I could hardly lift my feet any more. I knew that I couldn't stand it but a minute or two longer." The farmers were right. When the body sleeps, its core temperature drops, metabolism slows, heat escapes more quickly from the surface, shivering ceases—all of which hastens the loss of the dwindling supply of bodily warmth.

Johann Kaufmann and Peter Graber undoubtedly knew that sleep meant death, and very likely the younger boys did, too. They would have kept moving as long as they could—anything to avoid lying down in the snow. By late afternoon they had been wandering for three to four hours. They had come to the end of their endurance. Their cheeks and eyelids were raw from scraping the ice off their eyes. Their limbs were stiff. Their hands and feet and ears were beyond warming. Johann and Peter were no longer able to carry the younger boys, and Heinrich and Elias were barely dragging themselves through the snow. Every few steps they fell to their knees, struggled to stand, fell again. The older boys waited, turned back to help, shouted, fell down themselves. When they tried to think, words and images drowned inside their skulls before they could break the surface of consciousness. When they spoke they made no sense, to themselves or to each other. With body temperatures at 80 degrees, delirium entirely eclipsed reason. In the profound darkness of night in a storm, hope was impossible.

The younger boys gave out first. At some point Heinrich or Elias fell and couldn't get up again. Nothing Peter or Johann said or did could stir them from their apathy. It's possible the younger boys fell to their knees and the older boys, thinking they were praying, knelt beside them. That's how their families preferred to imagine it. In truth, whether they knelt together or not, it was a tribute to the strength of the bonds between them—and to the heroic efforts of Peter and Johann—that they had remained together at all. Other groups of children caught out in the storm straggled, separated, and dropped alone, one by one. But the five Schweizer boys hung together until the end. And perhaps they did pray with some small, glowing ember of the mind.

By all accounts they were good boys, obedient, cherished by their families. They had been prayed over and instructed in prayer from the moment they were born. Johann Kaufmann, the only one of Anna Kaufmann's first four sons to survive childhood. His younger brothers Heinrich and Elias, born after so much desperate suffering in the New World when their mother hardly dared to hope anymore. Johann Albrecht, the baby born on board the immigrant ship just three days out of New York Harbor—a child blessedly impervious to the noise and danger and heat of the long journey, and to the deprivations of the hard seasons ahead. Peter Graber, not yet four when his mother died of exhaustion and illness that first bitter spring in Dakota and raised to the age of sixteen by his stepmother in a house packed with brothers and sisters.

In the small, closed world these boys had grown up in, the only world they ever knew, the impulse to pray was so powerful and deeply ingrained that it was almost an instinct. And kneeling, if one or more of them had indeed summoned the will to kneel, would have been a blessing in its own right. Folding the body in the middle and bringing the knees close to the chest would have provided some protection.

It was better than lying in the snow.

❧ ❧

But in the end that's what they did. One by one, they collapsed onto the frozen ground. With his last surge of will, Johann raised himself and wrapped Elias in his arms. As the snow conducted heat away from their bodies, their heartbeats slowed to an occasional twitch. The boys lost consciousness. Beyond both hope and fear, they felt nothing at all.

❧ ❧

Doctors have a brutal phrase they use in treating people found unconscious in the cold: "You're not dead until you're warm and dead." In profound hypothermia, the internal functions become so slow and feeble that the body enters a kind of suspended animation. The pulse is all but gone, the brain barely flickers with activity, the blood moves glacially through the veins and arteries—but there is a window during which an unconscious hypothermia victim can be resuscitated with surprisingly little damage. A fairly wide window, in fact.

Had a rescue team found the five Schweizer boys and moved them exceedingly gently out of the storm and rewarmed them carefully with hot water or heated air, they might still have recovered. Assuming they slipped into unconsciousness around 4 P.M., the boys may still have been alive at 7 P.M., when the temperature had dropped to 10 below zero and the wind out of the northwest blew at 40 miles an hour. It's conceivable that one or more of them might have been resuscitated as late as 9 P.M.

But nobody found Johann, Heinrich, and Elias Kaufmann, Peter Graber, or Johann Albrecht—not in time to save their lives.

Once their core temperatures fell below 84 degrees, their hearts were beating at less than half the normal rate. With every degree of temperature loss, the heartbeat slowed and weakened. Just as their leg muscles had failed to obey commands to contract and release while they were still able to walk, so their hearts now became less and less responsive to the electrical signals transmitted by their nerves. Fiber by fiber, the cold was paralyzing their hearts. Eventually the signals were so faint that they failed to trigger any cardiac response at all. Circulation ceased. With no oxygen the brain guttered and went dark.

The boys lay on their sides with their arms pulled in tight and crossed on their chests and their knees drawn to their stomachs. By every vital sign, they were dead. They had no pulse, they were not breathing, their eyes were dilated, their brains were void of electrical activity. They were dead, but still they were not entirely gone. The cold that killed them also preserved the possibility of salvation. At normal body temperatures, the brain suffers irreversible damage three minutes after the heart stops beating. But in cases of profound hypothermia, the brain is so cold that it remains intact far longer. Modern doctors have succeeded in resuscitating an individual pulled out of icy water sixty-six minutes after cardiac arrest—full recovery with no brain damage.

Modern doctors.

Even the kindest, wisest rescuer in 1888 would have inadvertently killed the boys.

Before Thursday, January 12, 1888, ended at midnight and Friday the thirteenth began, every bit of moisture in the five young bodies, every cell, every tissue was frozen solid.

We'll never know how many spent that night out on the prairie. It had to be at least several thousand, most of them in the southern and eastern parts of Dakota Territory, in the eastern half of Nebraska, and in southwestern Minnesota. Northern Dakota was largely spared because the storm blew through so early that people remained home and kept their children in. Iowa, though it received the heaviest snow, also suffered relatively few casualties. The storm didn't hit there until late in the day, when evening was gathering and farmers and their children were back home. But in southern Dakota and Nebraska the timing could not have been worse. Sergeant Glenn estimated that 1 percent of "those overtaken and bewildered by the storm perished" and that of the dead 20 percent were children.

The catalog of their suffering is terrible. They froze alone or with their parents or perished in frantic, hopeless pursuit of loved ones. They died with the frozen bloody skin torn from their faces, where they had clawed off the mask of ice again and again. Some died within hours of getting lost; some lived through the night and died before first light. They were found standing waist deep in drifts with their hands frozen to barbed-wire fences, clutching at straw piles, buried under overturned wagons, on their backs, facedown on the snow with their arms outstretched as if trying to crawl. Mothers died sitting up with their children around them in fireless houses when the hay or coal or bits of furniture were exhausted and they were too weak or too frightened to go for more.

A young Dutch couple in Minnesota died kneeling side by side with their hands held high above their heads.

A nine-year-old Nebraska boy named Roman Hytrek was walking the prairie with his dog when the storm overtook them. That evening the dog turned up scratching at the door of a neighbor's house. Roman's empty coat was found in March. Eventually a search party recovered the boy's body. Roman had died alone leaning against the side of a hill. They speculated that he had unbuttoned his coat so that he could cradle his dog next to him in it and that the wind ripped it from his shoulders. But it may have been an instance of paradoxical undressing.

William Klemp, a newly married Dakotan in the full vigor of young manhood, left his pregnant wife at home and went out in the storm to care for their livestock. He never returned. A few weeks later, Klemp's wife gave birth to a son. It was spring when they found his body in a sod shanty a mile from the house. Klemp's face had been eaten away by mice and gophers.

In the region that would soon become the state of South Dakota there were deaths in thirty-two of the forty-four counties east of the Missouri River. Every pioneer who wrote a memoir, every family that recorded its history included a story of someone who died in the blizzard. Every story is heartbreaking.

Lois Royce, a young teacher of a Nebraska country school, huddled on the open prairie all night with three of her pupils—two nine-year-old boys and a six-year-old girl. The children cried themselves to sleep. Lois stretched out on the ground, lying on her side with her back to the wind and the children cradled in the hollow of her body. She covered their sleeping bodies with her cloak. The boys died first. Lois felt one of the bodies cease to breathe and go cold. Then, a few hours later, the other. The boys went in silence. The little girl, Hattie Rosberg, had begged her teacher through the night for more covers to keep her warm. She died at daybreak deliriously crying, "I'm so cold, mama, please cover me up." When the air had cleared enough to see, Lois left the three dead children lying together and crawled on her hands and knees a quarter of a mile to the nearest farmhouse.

In Dakota's Beadle County, six miles southwest of Huron, where Sergeant Glenn staffed the Signal Corps observing station, Robert Chambers, a farmer in his early thirties, was outside watering cattle with his two sons and their Newfoundland dog when the weather turned. The older boy, who was eleven, suffered from rheumatism, so Chambers sent him home before the storm got bad. He thought that he and nine-year-old Johnny could drive the cattle to the barn themselves. The dog would know the way. But, in Sergeant Glenn's words, father and son were overtaken and bewildered. When Chambers realized there was no hope of finding their farmhouse, he burrowed into a drift, wrapped Johnny in his jacket and vest (neither of them had come out with overcoats), and told the boy to get into the hollow out of the wind. Robert Chambers stood in the storm shouting for help as long as his voice held. The dog barked frantically. But no one heard them over the wind.

By evening Chambers was too cold to do anything but lie down in the snow next to his son. He put the dog beside them for extra warmth. Johnny could feel how frigid his father's body was. He urged his father to get up and to look for the line of the trees they had planted by the house. But Chambers would not leave his son.

As the night wore on, father and son talked about death. Chambers assured Johnny that they would survive and repeated over and over that the boy must lie still. Johnny knew that his father was freezing to death. At some point the boy dozed off. When he woke, his father was still alive, but barely. Chambers told his son to pray and that he would pray with him.

At daylight a rescue party heard the Newfoundland barking and found them. The snow had drifted so deeply that Johnny was entirely buried but for a small opening by his mouth. The dog was standing guard. Robert Chambers was dead.

The Westphalen girls, Eda and Matilda, also died in the night. Though born five years apart, the daughters of German immigrants, the girls had grown close to each other in the tragedies that had befallen their family during the past few years. Diphtheria struck the Westphalens in the winter of 1883. Two days before Christmas, six-year-old Frederick died. Six weeks later, their father, Peter, deranged by grief, hanged himself. Since then their mother had managed alone with six children. The winter of the blizzard, Eda was thirteen, Matilda, eight. The storm hit when the girls were at their country school in a hilly section of eastern Nebraska near the railroad town of Scribner (named by an Eastern railroad official for his son-in-law, New York publisher Charles Scribner). The teacher, Nellie Forsythe, told the children to go home. Eda and Matilda left together. The schoolhouse was halfway up the side of a smooth rounded hill; their house was a mile due north at the bottom of a valley cut by a creek. Usually it was an easy walk downhill across the fields. But in the storm the girls had the wind in their faces. No matter how they struggled against it, the northwest wind pushed them east into a series of ravines. For a while they wandered in circles. Then they drifted east and south with the wind. Only when they came to a wire fence did Eda realize they had gone in the wrong direction. They needed to turn around—but turning meant walking into the wind. Matilda failed, and Eda took off her wraps and covered her younger sister.

Most victims of hypothermia curl up on their sides and die in a fetal position. Eda and Matilda died facedown. Very likely they dropped while fighting to walk into the wind. Once they fell, they must have lost consciousness very quickly. They lay on the snow a few feet apart on the side of a hill. The windward side. All night the wind blew snow over their bodies, covering them and laying them bare again.

In the course of the night, the haystack in which Etta Shattuck had taken refuge became her prison. The hay had become so compacted and heavy with drifting snow that it pinned Etta in the small hollow she had dug for herself. As the temperature plunged, the fibers tightened. Etta's torso stayed fairly warm, but the cave was so shallow that she was unable to shelter her legs or feet. Exposed to the cold, her legs turned to blocks of wood. She was powerless to escape.

Etta drifted in and out of consciousness, but she never fell into a deep sleep. She felt mice rustling through the stack and nibbling at her wrists and somehow that comforted her. It seemed miraculous that something else was alive in the storm. When she was most alert, Etta prayed. She moved her lips and tried to summon the voice to sing hymns. She ran the words through her mind, but the sound that came from her mouth was hardly more than labored breathing. She was glad as never before that she had found God. God had brought her to the haystack; she was sure of it. God would guide the steps of a rescuer. Etta had faith. She knew she would be saved.

At some point in the night the wind died down enough for her to hear coyotes howling. That keening yelp. Or maybe it was still the wind. Etta's eyes fluttered open and the air looked a little brighter. It must be morning. Whoever had forked the prairie grass into this stack would come. Etta tried her voice to see if she could cry out for help. She could move her mouth and neck and shoulders. But her body was caught in the vise of the frozen haystack and her legs were paralyzed. The hymns and prayers would keep her going until someone came and pulled her out.

If nothing else, as long as she could sing and pray, Etta could ward off deep sleep—the sleep from which she would never rouse herself.

CHAPTER NINE

Prairie Dawn

Weather goes on forever with no direction or resolution, but a storm, like a story, has a beginning, a middle, and an end. The conditions that made the storm will in time unmake it. The seeds of destruction are present from the start. It is the nature of raging low-pressure systems to drag down calm, stable highs. Winds must relax as temperature and pressure gradients dissipate. Once the front moves through, cold air deepens. The clash of contrasting air masses continues—but somewhere else. Another story. On yesterday's battleground falls the hush of equilibrium. Air aloft dries as it spreads and sinks. Clouds break up. After the worst storms, the most beautiful weather often shines down on the scene of devastation.

Before dawn on Friday, January 13, 1888, the blizzard had pretty much blown itself out over the Dakotas, Nebraska, and southwestern Minnesota. The last gusts put the final touches on drifts and hollows, and then the atmosphere subsided in a deep sigh of high pressure. Over a thousand miles of prairie, from central Canada down to Oklahoma, the air grew pure and dense and dry with intense cold. Twenty-nine below zero at 6 A.M. at Fort Assinniboine. Twenty-five below at Huron. Seventeen below at Omaha. Sunrise two hours later barely budged the mercury. As the light came up, the dome of the sky seemed to lift and expand like a balloon filling with air. The colors of the most delicate alpine flowers flushed the sky from east to west—first grayish pink, then powder blue, then azure. The last bits of moisture condensed and froze and fell glittering in tiny crystals from the cloudless sky—diamond dust, meteorologists call it, a sign that the coldest, driest weather is building in.

"Silent as a marble sea and flaming with sunlight," Hamlin Garland wrote of the prairie the day after a winter storm.

In hilly Buffalo County, bordering the Missouri River in southern Dakota, Thomas Pirnie stepped out of his farmhouse on the morning of the thirteenth to survey a world transformed. "It was a beautiful but awe-inspiring scene," he marveled. "The frost sparkled like myriads [sic] of diamonds and the sun dogs were beautiful as a rainbow. Overall, there was a death-like stillness, not the sound of a dog barking, a cow bellowing or a horse neighing. The hills which had been sharply outlined were now but rounded knolls. Ravines had almost disappeared. Everywhere there was perfect whiteness. The smoke from the house chimneys went straight up in round columns high into the sky. This was the only sign of life about us."

The towns and cities were paralyzed, their wide, straight main streets drifted over, the shops and schools closed, the rail yards deserted. "Not a team or vehicle of any kind is to be seen," wrote a resident of Bismarck. "It reminds one of the ancient depopulated cities. The terrible swirling masses of snow prevents [sic] objects from being seen across the street." Bismarck's Signal Corps observer Sergeant Sherwood reported drifts as high as twenty feet. Trains were abandoned on impassable tracks. Rail service had ceased altogether on the smaller lines. Locomotives equipped with plows set out from Aberdeen, Sioux Falls, Omaha, Lincoln along the major routes but were unable to break through the wind-packed drifts. Heavy snow brought down telegraph wires. Not a word came or went from most of western Iowa. "Travel almost entirely suspended," wrote the Signal Corps observer in Des Moines.

The prairie was void of motion and color. Every object large enough to raise a profile had an amphitheater of white carved in its lee. The exposed northern and western faces were scoured clean by the wind. The first rays of sun spread fleeting colors but no warmth. The snow turned pinky yellow, the shadows blue. At intervals in the vast, smooth white surface bits of black and brown caught the light, dark irregular specks, as if the wind had snatched scraps of cloth from the farmhouses and barns and littered the open ground with them.

These were the cattle that farmers had failed to bring to shelter the day before.

Some lay on their sides completely encased in ice. Some stood with their heads anchored to the ground by globes of ice as large as bushel baskets that had formed from their congealed and frozen breath. In western Minnesota's Pipestone County, a Scot named Jackson rode out in search of the Black Angus herd that he had seen vanish into the blizzard while he had been out watering them. A few miles from home he came upon the first of the cows, dead in the snow. He rode on. The line of frozen cattle extended for ten miles, running from northwest to southeast, just as the wind had blown. "Most of them were frozen stiff," wrote David Maxwell Fyffe, a neighbor of Jackson's and a fellow Scot, "but sometimes they found one that had life in it tho badly frozen and they shovelled snow on it and passed on. They found out that covering with snow was wrong when they came back as the snow had started up a heat in the animal and it began to revive and commenced to struggle out of the snow and it got worse frozen than ever. They hauled a few of these home to the barn, but when they got properly thawed out the frozen flesh came off them in chunks. . . . Jackson was left with only two bulls which were kept around the barns."

❧ ❧

As the light grew stronger, smaller and finer objects became distinct in the vast whiteness. The gray sleeve of a jacket, the frozen folds of an apron, a boot, a shock of hair, a child's naked hand.

It seems incredible that a person could survive that night without shelter or food or warm clothing, yet most did.

Though some survived the night only to die instantly when they rose to their feet in the new light of day.

Twenty-five miles south of where Etta Shattuck lay imprisoned in a haystack, a blanket lay half buried in the snow. When the rising sun set the snow ablaze, hands pushed the blanket away and two children rose slowly from the ground. Amelia Shirk and Omar Gibson had lived through the night because of this horse blanket. The morning before, Omar and Amelia had been lucky enough to have a horse to ride to their country school. When the storm hit Holt County, their teacher, John Schneider, dismissed school and told Amelia and Omar to set out for home on their own. He was confident that the horse would get them back safely. But the horse went astray and the children were soon blind and lost. They dismounted, removed the horse's saddle blanket, and turned the animal loose. About three miles from their school, just over the Wheeler County line, they became exhausted and collapsed together in a snowbank.

Amelia was twelve, Omar sixteen, the younger brother of her stepfather. They weren't blood relatives. Nonetheless Omar looked after the girl tenderly. He wrapped her in the blanket, and when she wouldn't stop shivering he gave her his jacket as well.

They were both alive at sunrise. In fact, Omar was strong enough to get to his feet without help. He told Amelia he was going to search for the horse. Omar set out, took a few slow steps, and then Amelia saw him drop. He died moments later.

The same thing happened to a young boy named Jesse Beadel in Dakota's Jerauld County. Jesse and his grandmother had ridden together to his school the day before in a horse-drawn sleigh. The grandmother was filling in as substitute teacher for a couple of days. The blizzard struck late in the morning. At first the grandmother decided they must wait out the storm in the schoolhouse, but as the blizzard raged on, she and Jesse took pity on the horse, still hitched to the sleigh at the south end of the building. The grandmother's house was only three miles from the school, to the southeast, which meant they'd have the wind at their backs. They got most of the way there when the horse floundered in deep snow. Jesse unhitched the horse and freed it from the drift, but the sleigh was stuck fast. Even had they been able to walk through the deep drifts, Jesse and his grandmother had no idea where to go. The best he could do was tip the sleigh up on its side so it would break the wind. He gave his grandmother the robes they had brought along in the sleigh and huddled down beside her.

They endured the night. At dawn Jesse saw a house about half a mile away and he told his grandmother to wait for him in the shelter of the sleigh while he went for help. He pulled himself to his feet. He spotted the horse standing and looking at him across the snow just a few yards away. Jesse managed to stagger most of the way to the house before he collapsed. He, too, died in moments.

Jesse and Omar both died of cardiac arrest caused by shock. All night long as they lay in the snow, their bodies had fought to keep warm by drawing the blood away from their skin and exposed extremities and pushing it deep into the core. In the morning, when the boys stood up and began to walk for the first time in hours, the sudden change of position and unaccustomed motion triggered a massive fall in blood pressure. Blood from the cores of their bodies cooled as it moved into the cold extremities. Their pumping hearts forced this chilled blood to circulate. When it reached their hearts, the boys instantly went into ventricular fibrillation. Instead of contracting and releasing in steady rhythmic beats, the lower chambers of their hearts began to quiver without coordination or effect. Pumping quit; blood stopped circulating through their bodies. Jesse and Omar blacked out. Ten or fifteen minutes later, their fibrillating hearts ceased to beat altogether and the boys died.

The cold heart is extraordinarily sensitive. Rough handling, the mild jarring of feet hitting the ground, or simply the change of position from prone to upright can set off ventricular fibrillation. Coming in out of the cold can also stop the heart. In severe hypothermia, death by rescue is all too common. In 1979, sixteen Danish fishermen shipwrecked in the frigid waters off Greenland were rescued after more than an hour in the icy sea. All of them were conscious and able to board the rescue ship. They were wrapped in blankets and given warm drinks. When they stood to walk across the deck of the ship, all sixteen of them dropped dead. Their hearts, like Jesse and Omar's, had gone into ventricular fibrillation. Doctors call this rewarming shock.

That was what killed eighteen-year-old Frederick Milbier. Frederick had been caught out in the storm in an open bobsled with his sister, Christina, her husband, Jacob Kurtz, and their baby. They were on their way home from dinner at the home of Jacob's parents near Yankton in southeastern Dakota, just north of the Missouri River. When the horses refused to walk into the wind, Jacob left his wife, child, and brother-in-law in the bobsled and went for help. He made it only a few yards before the wind knocked him over. Unable to get up, Jacob lay on the snow and lapsed into a stupor. In the teeth of the storm, Christina Kurtz unbuttoned her dress and her blouse and placed her baby next to her naked skin. This probably saved both of their lives. In the clear light of morning,

Christina and Frederick saw a farmer in the distance standing by his haystack. Frederick got down from the bobsled, but, unable to walk, he had to crawl to reach the farmer. The farmer's family quickly got Frederick and his sister and the baby inside. Frederick died soon afterward in the warmth of the farmer's house.

Rescuers went out after Jacob Kurtz and found him buried in the snow. They reported that he was entirely frozen but for a sphere of warmth surrounding his heart. It's likely that Jacob Kurtz could have been revived in a modern hospital. But the search party that found him assumed he was beyond hope. Kurtz was brought to the farmhouse. Like his brother-in-law, he probably died of heart failure as he thawed.

When the light bored into her eyelids, Lena Woebbecke stirred. All through the night long, as the wind threw snow at her, she had drifted in and out of consciousness. Her brain was so sluggish with cold she could barely remember or comprehend what had happened to her. The storm rattling the windows of the schoolhouse, the children shoving their way to the door and the shouting in English she only partly understood, the quarrel with the boy about which way to go. She remembered leaving the boy and setting out by herself across the field. And the snow in her face. And turning back when she lost her way. But everything else was a blank.

Snow had melted under her body where she lay and then frozen again. When Lena tried to get up, she heard the ice crackle. She was frozen to the ground. With an effort she managed to free herself. She stood. Everything was clear now. She could see smoke from the frail chimney of the Woebbecke house. The school was just a few hundred feet behind her at the top of the hill.

Lena could manage only a few steps before she fell. She had no sensation in her feet or legs, and the drifts were deep. But somehow she was able to drag herself through the snow. It was downhill most of the way home, but at the bottom of the hill she'd have to cross a steep creek-bed and climb up the other side. Lena was near the creek-bed when her strength gave out. It was too cold. Exertion failed to warm her. Being female and big for her age was some advantage: a thin boy in her place would probably have been dead by now. But Lena had used up her reserves. Again she collapsed in the snow and lay still. Lena waited patiently for death.

Even before sunrise Wilhelm Woebbecke decided it was bright enough to resume his search for Lena. Over his wife, Catherina's, vehement protest he had gone out in the storm the previous afternoon to bring the child home from school only to find the school-house shut up and empty. By that time the blizzard was so bad that Wilhelm was unable to find the path and he wandered lost for an hour. When he finally, miraculously, stumbled on the house, he agreed with his wife that it was too dangerous to search any longer. Either Lena had found shelter with the neighbors or she was dead. Wilhelm and Catherina had done the best they could, probably more than Lena's own mother would have done. They had treated her like their own daughter ever since her mother had brought her to live with them in the summer. But they wouldn't risk their lives. They had their own children to worry about.

Wilhelm had just started out on his horse when he spotted a dark shape on the hillside opposite the ravine. He shouted Lena's name and watched in amazement as the girl lifted herself off the snow and rose to her knees. Though she could not stand, she held her hands up to show that she still had her dinner pail and reader. Wilhelm floundered through the deep snow that filled the ravine. Lena was unable to speak or use her legs, but Wilhelm managed to put her arms around his neck and she held on as he carried and dragged her through the ravine.

It was eight o'clock, the hour of January sunrise, when Wilhelm and Lena reached the house. By the time Catherina had gotten the child's frozen clothing off and put her to bed, Lena was in a coma.

❧ ❧

Johann Albrecht Sr. left for the Rosefield Township school at first light. His wife, Maria, had packed a large pail of food for him to bring with him. Enough food for their son Johann, for the teacher, James Cotton, and for any other students who had spent the night in the schoolhouse. There was no question in their minds that their boy and the teacher were safe at school. Even Maria had stopped sighing.

Johann Albrecht reached the Rosefield school without any trouble. Mr. Cotton was there to greet him—but no one else. "Where is Johann?" he demanded. Albrecht's English was limited. Cotton spoke no German. But Albrecht understood soon enough that his son and the four other Schweizer boys were missing.

Johann Albrecht could not comprehend why the teacher was safe inside the school while the boys were lost outside on the prairie, but he wasted no more time trying to get Cotton to explain. The neighbors must be told what had happened. They must start to search at once while there was still hope.

❧ ❧

The sun rose on the Reverend S. F. Huntley, warming his feet at his fire in Harmony Township, Jerauld County, Dakota—warming his feet and wondering about his wife. As befit a Congregational minister, the Reverend Huntley was a tall, gentlemanly figure of a man with a calm, settled demeanor and a physique that more and more resembled a bowling pin. S.F.'s wife, Abi, was a minister, too—an ordained Quaker minister and a founding member of the Harmony Township Friends Meeting. The Reverend Huntley would be the first to say that his wife was a pious, selfless woman; all her life she had taught school and preached the word of God. She was a person others turned to in adversity. But that only made her want of faith the previous night all the more baffling and disconcerting. Abi had bowed her head and prayed with him when he clasped his hands and entrusted their children, Mary, Ernest, Mabelle, to God. Together, as night fell and still the children did not return from school, they prayed to God to watch over the little ones, to shelter them in the storm and see them safely home. S.F. sincerely believed that there was nothing more to be done. But still Abi would not resign herself. All through the night S.F. heard his wife tossing and fretting. By morning, she was haggard with fear.

The Reverend Huntley felt it was his duty to remain calm. Surely, he told his wife in his most comforting voice, the children's teacher, Miss Hunt, had taken them to her boarding place close by the school. No doubt they had passed a comfortable night in her more than capable care. It had probably been a wonderful lark to be away from home and in the company of other children.

So confident was he of the children's safety that S.F. spent the first frigid hours of daylight doing chores in the barn that he had been unable to attend to the evening before. That was enough to convince him that he had been wise to stay home the night before instead of risking his life on a wild rescue mission, as Abi had demanded. Even in the few minutes it took to walk back to the house from the barn S.F.'s feet had become insensate. Before he could set out to collect the children from Miss Hunt's boarding place, it was imperative to thaw his feet and make himself thoroughly warm.

Which was why S.F. was sitting by the fire with his feet up and his boots off when his neighbor Mr. Bartie came rushing in without so much as a knock on the door and blurted out excitedly, "You better be seeing after your children. They stayed in a straw stack last night."

S.F. could not have been more astonished if the man had raised a fist and struck him. "Are they alive?" he demanded. "I don't know," Mr. Bartie replied maddeningly, "I didn't hear any particular."

"How do you know that they were in a straw stack?" "Mr. Knieriem was over to Mr. Dingle's and told us," came the baffling reply. "What did Mr. Knieriem want? Did he come out to tell you?" "No," said Mr. Bartie, "he came after some beef's gall." "Were his children in the stack?" S.F. asked with rising alarm. "Yes, the teacher and all the scholars." "Why didn't you find out if they were alive?" The impossible Mr. Bartie merely shrugged. "As soon as Mr. Knieriem told us I put on my coat and came right over to let you know."

Abi, having listened to the entire exchange, was beside herself. S.F. could only comfort his wife with the reflection that the children could not *all* be dead or Mr. Knieriem would have no need for beef's gall. Then he got his feet back in his boots, dashed out for his horse with Mr. Bartie trailing after, and they were gone. Abi stayed home to keep the house warm and look after their four-year-old daughter.

※ ※

May Hunt had been calm and competent all night, but at the sight of the tall, somber, bearded figure of the Reverend S.F. Huntley standing inside the Hinners' farmhouse door and staring down at her and the children with such an expression of dismay, she broke down. Miss Hunt buried her face in her hands and wept—and soon all her students were sobbing with her. What made it more awful was that Miss Hunt belonged to the Reverend Huntley's church, as did the parents of little Frank and Addie Knieriem. What would people say in church on Sunday when they learned that it was she, Miss May Hunt, who was responsible for the fact that seven children had spent the night of the blizzard in a haystack? What would the Knieriems say when they found out that their precious daughter Addie could not walk—and very possibly would never walk again?

It was a long time before Miss Hunt was collected enough to tell the Reverend Huntley what had happened. With much prodding from the students she got through the account of the ordeal in the gully, the wandering blind and freezing in the storm, the providential discovery of the haystack, the extraordinary bravery of Fred Weeks in keeping guard at the mouth of the cave and climbing out every few hours to check on the progress of the storm. She made Fred himself tell how at four in the morning he went out of the cave yet again and this time the air seemed thinner and he looked up and saw stars overhead—and there, less than a hundred feet away, was the Hinners' farmhouse that they had sought so desperately the previous evening. Fred was too shy to say any more, so Miss Hunt told how heroic he was to stagger to the house on the frozen blocks of his feet and shout and pound on the door until Mr. Hinner came out.

But that was where Miss Hunt broke down again. It was too terrible to describe the rescue and what Addie Knieriem had endured—indeed was enduring still.

Bit by bit, the story came out. Fred and Mr. Hinner returned to the haystack as quickly as they could, bearing lanterns and piles of blankets. They called to Miss Hunt to bring the children forth to safety. At first, the smaller children were groggy and slow to react. They began to shiver uncontrollably as soon as they got outside the cave. Fred, despite the condition of his feet and hands, was everywhere lending a hand. In a few moments all of the children were out of the straw cave and wrapped in blankets or shawls and staggering as best they could to the glow in the farmhouse window.

All but Addie. In the excitement of rescue, no one noticed at first that there was something the matter with Addie Knieriem. She was unable to stand up and had to be pulled from the back of the cave. It was her feet. They had gotten wet when she tumbled blindly into the gully and the scarves came off her shoes. As she walked through the storm, the wetness rapidly chilled her feet. By the time Addie took refuge in the haystack, the water in her shoes and stockings had frozen solid. As she lay huddled in the cave through the long night, her feet touched the straw walls: bodily warmth bled constantly from her feet to the straw. Addie was too cramped and too exhausted to remove the shoes—and even if she had, there was no way of warming and drying her feet short of making a fire, which was out of the question in the storm. Addie's feet remained encased in the ice of frozen wool and leather all night. At some point the feet themselves turned to ice.

Finally, after Fred and Mr. Hinner carried her to the farmhouse, they got the child's shoes and stockings off. Miss Hunt was appalled. She had never seen human flesh look anything like this.

Frostbite, the irreversible freezing of living tissue, is the body's way of cutting its losses in severe cold and increasing the chances of survival. Warm heart and lungs and brain are more critical than warm ears or fingers or toes. So the blood retreats inward to the body's core and the extremities are sacrificed. Hands and feet are especially susceptible because they are so remote from the body's source of heat and because they have no large muscles to generate warmth by moving or shivering. In a blizzard, exposed fingers can become frostbitten in seconds, especially if wet or in contact with metal. Tight, constricting clothing or shoes and dehydration also increase the risk of frostbite. Addie met every one of these conditions. It's possible that Miss Hunt had wrapped and tied the scarves around the girl's shoes tightly enough to have hastened the freezing of her feet by cutting off the flow of blood.

When the skin temperature falls to between 22 and 24 degrees, the moisture in the tissues freezes. The fluid between the skin cells of Addie's feet froze first. Where there had been soft, pliable skin bathed in saline solution there was now a shallow formation of ice crystals. As the temperature dropped through the night, the crystalline network expanded. The growing ice crystals extracted water from the adjoining layers of cells and froze it into new crystals;

facets joined to facets in a hard brittle web. It's likely that during the minutes that Addie walked through the storm with wet feet, the freezing progressed rapidly enough to rupture the cell membranes. Not only the fluid between the cells but the cells themselves froze solid, damaging the tissues beyond the point of recovery. But even in the relative warmth of the haystack, the insidious, silent freezing progressed. The ice penetrated deeper, taking her toes, radiating out from her heels, spreading into the tendons and cartilage. As the ice solidified, the surrounding flesh became dehydrated. The loss of water, and the consequent rise in saline levels, killed even more cells. Horrifying as it was, the transformation of flesh to ice was almost completely painless. Once the temperature of her feet fell below 45 degrees, Addie felt nothing at all.

By the time they carried her to the Hinners' farmhouse and got her shoes and stockings off, Addie's feet looked like grayish purple marble. Fred Weeks's feet also were swollen and blotched with a waxy sheen.

The standard home remedy for frostbite in those days was to rub the frozen flesh with snow and then let it thaw gradually in warm water, and it's likely that this is what the Hinners did with Addie and Fred. The technique of rubbing with snow became widespread in Napoleon's army during the catastrophic Russian campaign of 1812 after soldiers suffered devastating burns and tissue damage while warming their frozen limbs at open fires. Rubbing with snow was indeed less harmful than scorching frozen flesh or exposing severely dehydrated, hypothermic bodies to roaring campfires, but it was still harmful. Doctors today treating Addie and Fred at a modern hospital would first give them painkillers to dull the agony of rewarming and then plunge their feet into a whirlpool bath of 104- to 107-degree water. While their feet were thawing, the children would be rehydrated with warm fluids, strengthened with a high-protein diet, and started on a course of antibiotics. But even in a modern hospital the odds of saving Addie's feet would still be low.

With the treatment she received at the Hinners' farmhouse, the odds were essentially nil. Rubbing with snow, though it initially dulled the pain of thawing, ultimately aggravated the tissue damage. And the pain, though delayed, was fierce. As the frozen flesh melted, Addie felt as if her feet had caught fire. Unbearable itchiness followed. But the full extent of the injury—and the brunt of the pain—did not surface for several days. In deep frostbite, the cells lining the capillaries and small veins rupture and the liquid component of blood leaks into the surrounding tissues. Drained of fluid, the remaining blood cells "sludge" in the vessels. In time the sludged blood starts to clot, and once that happens circulation is cut off entirely. The sludging and clotting of the blood is finally more destructive to the tissues than the formation of ice crystals. For Addie, it killed any hope of saving what had frozen.

Where the blood in her feet had ceased to flow, the skin soon blistered grotesquely and turned an inky, purplish black. Clear or bloody pink fluid wept from the blisters. Gangrene set in—the death and putrefaction of the blood-starved tissues. There was no recourse but amputation. One foot was saved, though all the toes had to be cut off. Addie's other foot was removed by a local doctor named A. M. Mathias. The anesthetic effects of ether and chloroform had been known since the 1840s, and the latter was used extensively as an anesthetic during the Civil War. It's likely that Dr. Mathias was familiar with these anesthetics, and he may well have read of the many deaths during surgery caused by chloroform, which was far more potent than ether. There is no record of whether Dr. Mathias himself knew how to administer inhalable anesthetics, especially when operating on a child. It's conceivable that Addie endured the amputation of her foot and toes with no anesthetic aside from an opium compound like morphine.

Fred Weeks was comparatively lucky. For days after the storm, fluid wept from the blisters on his feet and legs and the blackened flesh peeled away. The pain was so bad he couldn't bear to have clothing touching the skin. It's likely he ran a high fever. But eventually the wounds healed.

The Reverend Huntley's children, Mary, Ernest, and Mabelle, escaped with only swelling and blistering on their feet and shins. They clung to their mother and wept pitifully when their father brought them home later that morning. Abi Huntley wept, too, once she saw their feet and legs. After the children were fed and put to bed, Abi cornered her husband. She had never once complained since moving to Dakota Territory four and a half years ago—pregnant, homeless, well on in years, burdened with the care of three precious children. But this was past enduring, she told S.F. bitterly. They must pack up and go back east as soon as possible.

For most, the suspense of the night ended that morning—one way or another. In the clear light of day husbands tracked down wives who had wandered out into the storm. Dogs returned home, with or without their masters. Parents rushed to country schools where their children had spent the night next to fires of burning desks and chairs. Or the schools were empty, the children unaccounted for, the teacher dazed with grief and remorse. News of the living and the dead and the still missing got carried into towns on foot or horseback and radiated out again from the hotels and Western Union offices and railroad station agents' offices.

Story after terrible story circulated on the telegraph wires.

Mr. Stearns, a Dakota schoolteacher, had taken his three children to the school he taught near De Smet the day before and still had not returned home.

Frank Bambas, a Czech farmer, uncovered the body of his wife while shoveling a path from his farmhouse to his barn.

A Nebraska settler named Closs Blake came upon a bobsled turned upside down and buried under a snowdrift; when he brushed off the snow and righted the bobsled he found the body of a little boy frozen underneath.

In Turner County, south of Sioux Falls, Peter Wierenga enlisted neighbors to help him search for his four children. Together the men slowly walked the route the children took home from school. They spotted Wierenga's seventeen-year-old daughter first. She was in a grove of saplings with her back to a tree. She had frozen to death standing up. Her brother and two younger sisters were huddled at her feet. All four of them were dead.

Peter Heins, in the same county, lost three boys. Crazed with grief, Heins was on his way to the schoolhouse, yelling that he was going to kill the teacher, but a neighbor whose children had survived stopped him and told him what had happened. When the storm came up the teacher had begged the children to stay in school and she even locked the door. But the children refused to obey. One of them was seventeen and he led the rebellion. The kids overpowered the teacher and managed to get the door open and then they fled for home into the storm, Heins's three sons among them. The boys made it two miles before they collapsed in a pasture.

Other stories of that Friday morning after the storm came to light only later.

John Jensen, a handsome young immigrant from Denmark living in Dakota's Aurora County, not far from the Huntleys and the Knieriems, had gone out to his well to shovel snow when the storm hit around noon. He wandered, lost, for seven hours before he finally stumbled on a neighbor's house. When he got inside, the neighbors had to cut the clothes off his body, snow was so tightly packed and frozen into the fabric. Though safe himself, Jensen was desperate with anxiety at leaving his young family alone in their farmhouse—his wife, Nickoline, who had never wanted to move to Dakota in the first place, their daughter Alvilda, and their month-old baby, Anna Nickoline. A week later Jensen wrote a let-

ter in Danish to his sister describing the tragedy that had befallen
him:

At seven the next morning I started and thought I would
go home. When I came home I found my loving wife outside
the house, froze to death, and then I went into the house and
there lay Alvilda on the floor froze to death. After what I can
see from the tracks she [Nickoline] had gone to the stable. She
had many clothes on but that didn't help. The little one
[Anna Nickoline] in the wagon was living.

She was a very good wife and it makes me feel that I wish I
was dead. But it must be God's will that it should happen so.
But what I am glad over, we shall meet in HEAVEN, where
we will never part.

After discovering the bodies of his wife and daughter, Jensen
walked two miles to his brother Andrew's house with the baby in
his arms. When Andrew heard what happened he asked, "What
did you do, John?" "I cared for the living," was his reply.

✿ ✿

"A scene became quite familiar in many localities," a Dakota historian wrote of the immediate aftermath of the storm; "the arrival of
a party in quest of a doctor and bearing either on their arms or in
some sort of conveyance, the half frozen body of a neighbor or two
who had been exposed to the storm. . . . The heartrending cries of
the bereaved were heard in a hundred homes. . . . The whole scene
covering almost the entire territory remind[ed] one familiar with
battles of the scene around the hospital camps while a hotly con-
tested battle was in progress and the wounded were being borne to
the rear and turned into the surgeon's care."

✿ ✿

The Schweizer community came out in force once Johann Albrecht spread the word that the five boys were missing. It was bitter cold and clouds blew in by Friday afternoon, making it hard to see anything on the surface of the snow in the flat light. Around the schoolhouse there were many fresh tracks in the snow where parents had come and gone, so it was impossible to pick up a trail. The men made wide circles around the school, but they saw nothing. Not a trace of the boys. A light snow began to fall around three o'clock that afternoon and continued into the gloaming hours. Darkness came early. They would search again the next day.

South of Scribner, Nebraska, some seventy-five men had spent the day searching for the Westphalen sisters, Eda and Matilda. The men had gone out with poles and shovels and they were working methodically through the drifts. A few of them spotted tracks on the sheltered side of a hill, and someone else found deep indents in the snow. They speculated that the girls had stamped their feet here trying to warm themselves. Some men said they were sure that the tracks circled back on themselves. They followed every ripple in the surface of the snow. They dropped the poles into the deeper drifts, thinking they would feel the difference between a frozen body and frozen earth. But by dark, the searching parties gave up and went home. Someone was sent to the Westphalens' house to tell the widow that there was still no sign of her two daughters.

CHAPTER TEN

Sunday

In the days following the blizzard of January 12, 1888, the cold came down from the north like a river in flood. Channeled by the Rocky Mountains to the west but otherwise unimpeded, the frigid Canadian air mass surged south on the back of the intense temperature and pressure gradients that had unleashed the storm. Alberta, Montana, the Dakotas, Nebraska, Kansas—as long as there was energy to propel its flow, the cold air would keep moving south. Only the most robust cold fronts make it as far south as Oklahoma and Texas in howling blasts they call blue northers. The front that spawned the blizzard of January 12 was so robust that it traveled all the way down to Mexico, pushing cold air into Veracruz and through the narrow waist of land pinched between Oaxaca and Chiapas, until it finally issued into the Gulf of Tehuantepec.

Chief Signal Officer Greely was well aware of how North America's terrain acts to direct the flow of cold waves. As he

wrote in his book *American Weather,* published the same year as
the blizzard,

> There are very marked topographical features in the
> United States, which result in causing to advance from
> British America [in other words, Canada] the greater part of
> the winds following in the wake of cyclonic storms. The
> Rocky Mountain range, averaging about nine thousand feet
> in elevation, is as high or higher than the upper strata of most
> low-area storms, and so the air current cannot be drawn from
> the westward. Again, the broad, vast valley drained by the
> Mississippi descends with a gradual and substantially unbro-
> ken slope from British America to the Gulf of Mexico, so that
> any air flowing northward must be considerably retarded in
> its movement by the great friction arising from moving over
> continually ascending ground. On the other hand, the air
> from British America passes off gradually descending sur-
> faces, and this movement is further facilitated by the air
> being dry and cold, and hence dense, which naturally under-
> runs with readiness the lighter, warmer air of the retreating
> low area.

Though Greely's prose is clotted, his meaning is clear: the lay of
the land itself between Canada and Mexico fosters the southern advance of cold waves just as it hinders their movement to the north.
The cold wave of January 12 through 15 was a classic American
weather pattern, though amplified to an almost unprecedented extreme.

In fact, it was the rapid southern advance of the cold wave that
finally made General Greely sit up and take notice. The general
had spent the day of the storm dealing with the usual bureaucratic
vexations attendant on his position: a scrawny young man named
Edwin R. Morrow wanted permission to enlist in the Signal Corps

despite the fact that he was two inches under height and eleven pounds underweight; one of the seventeen clerks who worked at the Washington, D.C., Signal Office was taken to task for racking up twenty-nine sick days in 1887 and received a stern warning not to "cause any loss of service to the government during the current year"; a couple of new details emerged in the New Orleans scandal involving Signal Corps observers who had been squeezing the local cotton exchange for cash payments for weather data; Lieutenant Woodruff in Saint Paul requested and was granted access to weather reports from Canadian stations in Minnedosa, Swift Current, and Calgary.

On Friday the thirteenth, Greely undoubtedly saw telegrams about the extent and severity of the storm from both Signal Corps observers and railroad agents, and the first accounts of the suffering and casualties in the Dakotas and Nebraska appeared in the Saturday morning papers. But it wasn't until Saturday night, when the cold wave was on the doorstep of Texas and Louisiana, that the general finally realized that the situation warranted special attention. The threat to Southern sugar growers was what alarmed him. On Sunday morning, Greely broke with his usual custom and reported to his office at the Signal Corps' recently acquired headquarters at 24th and M Streets. As he scanned the daily weather maps and the outgoing telegrams of warnings, his alarm turned to outrage. Even a fool could see at a glance that cold air was barreling south like an express train. Just look at the observations from midnight on Saturday—Palestine (southeast of Dallas) was already reporting 24 degrees with strong northerly winds, and San Antonio, farther south and west, was down to 46, while the temperature at Galveston on the coast was 72. The pressure gradient across the state was practically unheard of. Why in God's name had that puppy, Junior Professor Henry A. Hazen, delayed sending out the signal to hoist the cold wave flags in Texas and Louisiana? It was perfectly obvious that the warnings should have gone out the day before in the 3 P.M.

dispatch or at the very latest at 11 at night. But to wait until the freezing air was hours—indeed, minutes—away from thousands of acres of precious sugar and cotton plantations was as baffling as it was insupportable.

Bent on limiting the damage, Greely took charge at once. He fired off telegrams to New Orleans stating with grave emphasis the severity of the cold that would be upon the sugar districts by that night or the first hours of Monday. And then he set about boxing young Hazen's ears. "Your attention is called to the fact that errors of such a kind as that made by you seriously injure this service in the minds of those reading such a dispatch," Greely dictated, fury getting the better of his prose. "It is hoped that you will exercise such care in this respect in the future with these dispatches as will avoid a repetition of such carelessness." Warming to his subject, Greely decided to use this opportunity to slap Hazen down for another transgression. Back in October, when he was spanking new to his position as civilian forecaster with the Signal Corps, Junior Professor Hazen had engaged in an unseemly public contest with a meteorologist at Boston's venerable Blue Hill Observatory to see which one issued the more accurate forecasts for a month of Bostonian weather. Hazen then had the gall to publish the results in the December 30, 1887, issue of *Science*. (Naturally he claimed to have bested the Blue Hill forecaster; just as naturally the Blue Hill forecaster, one H. Helm Clayton, argued in a rebuttal published two weeks later that Hazen had rigged the contest with fuzzy terminology. Clayton insisted that if one defined as "fair" a day on which less than .01 of an inch of rain fell and as "foul" a day with .01 of an inch or more of rain, then *his* forecasts were far superior to Hazen's.) Greely ignored this definitional hairsplitting and cut to the heart of the matter: He felt "great dissatisfaction" that Hazen had availed himself of "the first opportunity of an important public duty for advancing your personal interests or gratifying a desire to settle a question of personal standing professionally between Mr. H. Helm Clayton and yourself."

There was more, a great deal more, that General Greely would hold Junior Professor Hazen accountable for—but this would have to suffice for now. The general knew full well that it was he, not Hazen, who would have to answer to the irate sugar planters of Louisiana, not to mention the jackals of the press who never passed up an opportunity for taking the Signal Corps to task. Indeed, just the day before, the *New York Tribune* had run a prominent article under the glaring three-tier headline "Weather Prophets at Odds, A Ludicrous Blunder Yesterday, A Signal Service Inspector Says More Gumption and Less Science Are Needed" in which a reporter wrote with obvious relish that the Signal Corps had issued a forecast for "warmer, fair" weather when in fact the day turned out just the opposite: "The manner in which the prophecy was not fulfilled was remarkable. Seldom have innocent and trustful people been worse beguiled than New-Yorkers who arose in the morning and beheld, not blue skies and clean streets, but rain pouring down upon snow four inches deep, rapidly converting it into slush and a thick haze of fog enveloping earth and sky, and making matters still more unpleasant." What was still more unpleasant for Greely was that one of his indications officers, the excitable Irishman Lieutenant John C. Walshe (the same Walshe who had inspected the Saint Paul office in December and gotten embroiled in Woodruff's dispute with Professor Payne and the Chamber of Commerce), had blabbed to the *Tribune* reporter that the reason government forecasts were so bad was that "the man now in charge of that branch of the service at Washington [Professor Cleveland Abbe] is too much of a scientist and too little of a weather observer." Walshe rambled on that "Unless a man spends a long apprenticeship learning the details of weather conditions all over the United States he will fail in making predictions no matter how good a scientist he may be. The general rule that will apply in one place won't apply in another. A prediction like that sent for to-day does more harm than one might suppose."

More harm indeed. Lieutenant Walshe would very shortly be hearing from the general as well.

※ ※

In Saint Paul, Lieutenant Woodruff had Sunday off—actually, just Sunday morning and afternoon. In the evening he'd have to climb once again to the top of the Chamber of Commerce building and spend a few hours studying the maps and incoming telegrams so he could issue the midnight forecast for Monday.

But Woodruff did not have to be in his office surrounded by a litter of maps and telegrams to know that the cold weather that had set in since Thursday's storm had reached a climax, or rather a nadir, that Sunday morning. He could feel it himself pressing in at every window and door in his residential hotel, cold that no amount of modern steam heat could keep at bay. Over the entire region that Woodruff had charge of, observers were recording the coldest weather of the season—in many locales the coldest temperatures ever measured. Thirty-six below zero at 6 A.M. Central Time according to Sergeant Lyons's thermometer on the Chamber of Commerce roof (six days later it would fall even lower, to an all-time record of 41 below). Forty-three below at nearby Fort Snelling. Thirty-seven below at Professor Payne's observatory in Northfield. Twenty below reported by Sergeant Glenn at Huron (temperatures had bottomed out there at 31 below the previous night at 9 P.M.). Twenty-four below at Yankton. Twenty-five below at Omaha. Thirty-five below at North Platte, Nebraska, the lowest temperature ever known in the history of the county, with a barometric pressure of 30.94, the station's highest pressure then on record.

Woodruff himself had recorded a temperature of 63 below zero on his spirit thermometer during the snow winter of 1880–81 while fighting the Sioux in Montana under Colonel Guido Ilges (the all-time record low for Montana, and the lower forty-eight, for that matter, is 70 below, recorded at mile-high Rogers Pass on January 20, 1954). But as far as personal comfort was concerned, Woodruff conceded that there really wasn't much difference between 63 below in Montana and 36 below in Saint Paul. At least the winds were calm that Sunday morning, and the sun shone feebly in a sky of steel.

The only other consolation was that this unprecedented cold wave was the ideal weather for building a rock-hard ice palace for the upcoming Winter Carnival, the high point of the winter social season at Saint Paul. Though Thursday's blizzard had forced city officials to postpone the ceremonial laying of the corner block, the occasion took place with all due pomp on Saturday evening—a bit of luck, as it turned out, since this just happened to be the day Woodruff celebrated his thirty-ninth birthday. When it was finished a few days hence, the ice palace would rise to 120 feet at its highest turret, with some 55,000 22-by-33-inch ice blocks covering an acre of ground and weighing in at 6,000 tons—all of it frozen solid by the subzero weather and glittering like diamonds.

Woodruff had already requested Greely's permission to conduct experiments on the effect of all that ice on the temperature of the surrounding air. He would need to requisition half a dozen maximum-minimum thermometers from the government to carry out the research. All for an excellent cause. With any luck, he and young Sergeant McAdie would publish their findings in the *Scientific American.*

Despite the terrible cold that Sunday morning, Anna and Johann Kaufmann went to *Gottesdienst* (worship service) at the stark white Salem Mennonite Church. They brought their three surviving children with them, six-year-old Julius, three-year-old Jonathan, and baby Emma, whose first birthday they had celebrated two weeks before on New Year's Eve, when they still had six children living in their house.

Maria and Johann Albrecht attended church as well. Bowed with grief, the couple looked barely bigger than their four young children. Maria sat with her hands crossed in her lap. Johann, at the age of forty-one, already had more hair in his beard than on his head. Now that their son Johann was gone, Peter at age nine was the oldest child. A thin long-faced boy with prominent ears and very dark hair, Peter looked terrified that his mother would burst into sobs in front of everyone. What a blessing that the child hadn't gone to school with his older brother that day. How many times had Peter heard his mother repeat this?

The Graber family was also at church. The two brothers, Johann and Andreas, stared fixedly at the ground rather than meet the eyes of the congregation. They were the only ones who had gotten home safely with Mr. Cotton when their brother Peter and the four other boys wandered off into the storm. No one could understand why God had chosen to spare these two and take the other five. Or why they were unable to find a trace of the missing boys even three days after the storm.

The Salem church was full and loud with talk before the service began. The names of the missing boys passed from mouth to mouth. Their fathers had been searching, all the able-bodied men of the community had been searching for two days now, and still there was nothing but some stray tracks that vanished in the drifts. Five sets of footprints in some places; three sets of tracks in other places.

The faithful prayed long and hard in those days. The Mennonite worship service was still in progress in the waning light of afternoon when the minister stopped to make an announcement. Afterward, when members of the congregation talked about it, no one could understand or explain why he had waited so long.

The minister told the faithful that one of their members named Johann Goertz had come forward with a startling discovery. That morning before church Goertz went walking out in his field beyond the fire break he had plowed around his house as a protection against prairie fires. About forty yards from the house, just along the southwest edge of the fire break, Goertz spotted something he had never seen before. It was an arm emerging from the snow. The arm was raised in the air as if in defiance or triumph. Clutched in the motionless fist was a cloth coat. Goertz approached and brushed the snow away. Before he left for church he had uncovered five frozen corpses. He couldn't tell who they were or what family they belonged to.

That was the minister's announcement.

The three fathers set out immediately. The mothers went home with the children. The Goertz farm was three miles due east of the church. The men got there as quickly as they could through the deep drifts. From a long way off they could see the saplings of the timber claim, black whips bent and twisted by the wind. The land shelved up in a steady rise ahead of them, like a long smooth wave pushing against the sky but never breaking. A bleak stretch of prairie.

The section lines made it easy to calculate distance and reckon direction. The fathers knew without thinking about it that their sons had wandered nearly three miles from their schoolhouse, three miles south and east. Like every living thing caught out in the storm, the boys had drifted with the wind and then fallen.

At first, the men stopped in horror when they saw the dark patches of cloth against the snow. Johann Kaufmann cried out, "O God, is it my fault or yours that I find my three boys frozen here like the beasts of the field?"

It was terrible beyond words to pry the children off the ground. At first it was impossible to separate the bodies, the boys had died so near each other. Johann Kaufmann had to carry his sons Johann and Elias together because the older boy died with his arms around his younger brother.

Night was falling by the time Johann returned home. Anna stood at the door with her three little blond children and stared as her husband carried the three bodies inside. The *freundliche* mother, small and soft-featured, who always had a smile for her children, watched without saying a word.

Johann set the rock-hard bodies on the floor next to the stove. Anna looked at her dead sons and began to laugh. She couldn't help herself. Her husband and her two little boys turned to her in disbelief but Anna didn't stop. It would be days before they could get the bodies into coffins. Anna laughed.

Emma was still a baby, too young to know what was happening, but Julius and Jonathan were old enough to understand that those frozen blocks next to the stove were the dead bodies of their brothers. For the rest of their lives, the two brothers would never forget the peals of their mother's agonized laughter.

❧ ❧

Sunday was nearly over by the time Daniel D. Murphy got out to his haystack. He almost hadn't bothered to bring in hay at all, there were so few animals left to feed. A bull and a team of horses were all that the blizzard had spared him. Fifteen years since Murphy had come to America from Ireland, first to work in the mines in Michigan and Montana, then to scrape together what he could and file on this ranch outside of O'Neill, Nebraska, and this was all he had left. A bull and a team of horses.

Murphy took his hired man with him to the haystack. He knew there would be a job shoveling off the snow piled up by the storm before they could even get to the hay. Cold as it was and getting dark, it was not a chore he relished.

Between the swish and crunch of shovels against snow and hay and their desultory conversation, Murphy and the hired man didn't hear the voice at first. When it finally grew loud enough to pierce their consciousness, the two men dropped their shovels and stopped dead to listen. There it was again, a thin quaver coming from inside the haystack: "Is that you, Mr. Murphy?"

The whole neighborhood had been out searching for Etta Shattuck for three days now. Why hadn't it occurred to anyone to look in Murphy's haystack?

Murphy and the hired man dug Etta out carefully. It was a miracle she was alive at all, after seventy-eight hours without shelter or food or water in the coldest weather ever known in Holt County. The girl murmured something about how Christ had covered her defenseless head, but Murphy urged her to be still and conserve her strength while they carried her to the house. His wife would do what she could to get the frost out of the poor girl's arms and legs. Rubbing the frozen parts with snow was the only hope, though after three days God only knew what condition she was in.

It was too late in the day and too cold to be starting for the Western Union office in O'Neill. Murphy would go Monday morning first thing. Etta's family in Seward must be notified. Murphy could only hope to God that she lived that long.

❦ ❦

All day Sunday the cold streamed down into Texas and Louisiana. Rain changed to sleet and then snow. Standing water froze. The thin uninsulated walls of houses rattled in the gale. Cold wave warnings came too late or not at all.

At the Signal Corps station in Abilene, the signal flag could not be lowered on Sunday because sleet had frozen it to the flagpole.

Galveston on the Gulf coast got hit a few hours later. The Signal Corps observer wrote in the station journal that "the air was filled with fine drifting snow or freezing mist, which, owing to the influence of a wind of forty miles per hour, cut like drifting sand and covered everything with ice." The temperature plunged from 62 degrees to 29 in a matter of hours. For the first time in the history of the city, milk vendors failed to deliver their goods; horse-drawn hacks and hotel buses were abandoned in the streets. "The blizzard came upon the city with remarkable suddenness and without the slightest warning from the weather bureau at Washington," noted a local newspaper. "The bulletin ordering up signals and predicting freezing weather arrived only thirty minutes before the blizzard itself."

Corpus Christi was next. The temperature dropped from 55 to a low of 16 during Sunday, and light snow fell through the day. Wind speeds could not be recorded accurately at the Signal Corps station because the cups of the anemometer were coated with ice. No cold wave warning had been received from Washington. "Much suffering was caused by the extreme and sudden change of temperature," the local observer wrote in the station's journal, "and this service received several criticisms for failure to give warning of the expected cold wave and storm."

It was raining in New Orleans late Sunday morning, but at 2:15 P.M. the wind veered suddenly from southwest to northwest and the temperature began to plummet. Eventually the temperature would fall some 35 degrees, from almost 80 before the cold wave to 45 that night.

Brownsville, at the southern tip of Texas, had its coldest weather since December 31, 1880, and January 1, 1881, with temperatures falling nearly 40 degrees in eight hours. Ice an inch thick covered trees, houses, and fences and brought the Western Union telegraph lines down. "Not much damage to crops," the observer noted, "but cattle and people suffered greatly." For the first time since white settlement, the Colorado River in Texas froze over with ice a foot thick.

By nightfall on Sunday, ice had downed the telegraph and telephone lines connecting Galveston to the mainland and the city remained cut off for the next twenty-four hours.

❊ ❊

The Westphalen girls, Eda and Matilda, were not found until midday on Monday, January 16. They were lying on their faces on the snow, their frozen bodies mostly exposed. It was a mystery how scores of searchers could have combed the hills and ravines around the school and missed them lying there, plain as day, on the side of a hill on John Haun's farm. Two miles from their home, almost due east.

The only comfort the men could bring their mother was that Eda, the older child, had wrapped her little sister Matilda in her shawl. Thirteen and eight years old.

It was a slow, sad walk back to the Westphalen place. One of the men would have to come forward and ask the girls' widowed mother where she wanted them to set the bodies down.

❊ ❊

General Greely, though "very much crowded with work," as he complained to one of his many correspondents, found time in the course of that week to give Junior Professor Henry A. Hazen a thorough dressing-down for his ineptitude over the cold wave warning in the South. The general issued a severe blast on Wednesday: "It is not understood how you possibly delayed the display of this cold-wave signal until midday of Sunday when you already sent a special message that freezing temperatures would prevail that very morning, nor why, under similar circumstances, you neglected to display the cold-wave signal at Galveston." Hazen's delay, continued Greely, "tends to bring the service into ridicule as such a dispatch for Texas and Louisiana was evidently a saving clause and not in the interests of the people." As a direct result of Hazen's actions, "one of the most severe and pronounced" cold waves in a quarter century "reached the Texas coast practically unannounced."

A few days later, Greely upbraided Hazen again for his "disposition toward captiousness, irascibility [and] petulance," and warned the junior professor that in the future he must "prevent even a tendency in this direction."

And with that, the blizzard and the cold wave it ushered in dropped from official Signal Corps correspondence.

CHAPTER ELEVEN

Heroines

News of the storm reached the major cities first. The *Saint Paul Dispatch,* an afternoon daily, carried reports of the blizzard on the same day it hit, and the Friday papers in New York and Chicago were full of the storm. The *New York Times* reported on Friday that it had received word via telegrams moving east along the tracks of the Northern Pacific Railway that the worst storm in twenty years was raging in the Upper Midwest. "Very Severe in Dakota," noted the *Chicago Tribune* that day.

Chicago got the news before Dakota. In Dakota itself and throughout the blizzard region there was something of a news blackout. Though the major telegraph lines connecting big cities remained intact, local telegraph lines and the few existing telephone lines were down, and trains were at a standstill. No trains and no wires meant no hard news. For the first few days after the storm, editors in Dakota and Nebraska were reduced to looking out their windows at the drifts and improvising. The initial accounts in the local dailies were whimsical, ironic, folksy—the blizzard was just another bad trick that old man winter had pulled out of his sack.

It wasn't until Saturday that editors throughout the upper Midwest realized they had a major story on their hands. The chuckling abruptly ceased. Rivers of black boldface ran down the front pages of the papers in Omaha and Lincoln, Huron and Sioux Falls and Aberdeen. THE DIRE STORM AND ITS FRIGHTFUL RECORD. UNPRECEDENTED LOSS OF LIFE. MANY LITTLE CHILDREN PERISH WHILE ON THE WAY HOME FROM SCHOOL. In long unbroken columns the papers listed the names of the dead and missing. Every paragraph reported a fresh tragedy: Fred Eller, an Omaha cigar maker, found frozen to death within a block of his boardinghouse. Emil Gilbertson, formerly of Chicago, frozen to death eighty rods from John Murphy's place near Hitchcock, Dakota. George F. Allen overtaken by the storm when he went out with his boy to get a load of hay, resulting in the death of the boy and the freezing of the father's feet and arms. The story of the ultimate sacrifice that Robert Chambers made to keep his son Johnny alive and the valiant loyalty of their Newfoundland ran for days on the front page of the *Daily Huronite,* and eventually turned up in papers all over the country.

By the start of the new week, the tally of the dead was the lead headline in most of the major metropolitan papers. On Tuesday, January 17, the *New York Tribune* put the death toll at 145 and "growing almost every hour." The *Tribune* kept the story on its front page for a week, reporting 217 dead on Thursday, January 19, and 235 by Saturday, January 21. "THE RECORD OF THE DEAD" was the right-hand lead headline in the *Chicago Tribune* that Saturday. The next deck down promised "Thrilling Tales of Suffering from Exposure—Men and Women Perish on Their Own Premises—The Dire Distress of the Passengers on a Blockaded Train—Babies Frozen in the Arms of Their Mothers—Many Indians Missing."

Pure catastrophe is fatiguing. The city dailies soon wearied of tales about frozen children and blockaded trains. After only a few days, accounts of amputations, record cold temperatures, and staggering losses of livestock began to blur together. It was a Nebraska paper called the *Omaha Bee* that found a fresh angle to keep the story alive and tingling. On Wednesday, January 18, the *Bee* reported that one of its "representatives" had "learned an interesting tale of the pluck and good judgment exhibited by a young lady schoolteacher of Valley County." "A Heroine of the Storm," proclaimed the headline, and the story went on to describe how a nineteen-year-old teacher named Minnie Freeman had rescued her "little brood" of pupils after "a terrific gale, sweeping everything before it, struck the [schoolhouse] and carried away in the twinkling of an eye the entire roof of the structure, leaving the frightened little ones exposed to the elements." Two days later, the *Bee* was back with new and juicier details. Somehow its "representative" had gotten hold of a photograph of the heroic schoolteacher and the paper was now pleased to report that Miss Freeman was a very attractive young lady indeed, that she was engaged to be married to a commissioner from South Omaha (swiftly denied by the heroine herself), and that she evinced a becoming personal modesty in protesting that she had done nothing out of the ordinary in rescuing her little brood. The storm now had its first certified heroine.

The *Bee* did not let matters rest there. Noting that its story on Miss Freeman had "excited wide interest," the paper declared: "She has become a heroine, and deserves to be rewarded. In France she would be voted a life pension. It has been suggested that this brave young lady, aged only nineteen years, be given a medal. The *Bee* would make another suggestion. Miss Freeman deserves something more substantial than a mere souvenir. She is now earning a scanty livelihood at probably $25 per month. She should be liberally rewarded by contributors in money that would enable her to acquire a home and become independent."

Meanwhile, a second heroine had come to light—Miss Lois (or Louise or Loie, depending on the whim of the reporter or copy editor) Royce of Plainview, the young teacher whose three small students died in her arms the night of the blizzard. Lois herself survived, but now faced the imminent loss of both of her feet and possibly one of her hands to severe frostbite. She, too, the *Bee* argued, deserved to be recognized and rewarded.

And so was born the Heroine Fund. Each day the *Bee* solicited—indeed begged—readers to contribute what they could to reward these paragons of young womanhood. Tallies of daily receipts preempted the death toll figures. Interspersed with lists of contributors, which the *Bee* soon took to calling The Roll of Honor, were snippets about the sterling lives and characters of the heroines themselves. Minnie Freeman took center stage as the Gibson girl, all apple cheeks and chin-up, commonsense bravery, while Lois Royce lurked in the shadows as the languishing heroine of a penny dreadful.

Then, before interest flagged or attention strayed, fate handed the press a third heroine. From a journalistic point of view, the third was the best of all. Like Minnie and Lois, this new heroine was a teenage schoolteacher, but her tale was far more intriguing and, better yet, her fate still unresolved. It began with a mysterious disappearance into the white maelstrom; there followed a fruitless search, an agonizing brush with death, an eleventh-hour rescue; now there was a young life hanging in the balance as the likelihood of recovery from grievous injuries waxed and waned every hour. The papers had learned of the ordeal of Etta Shattuck.

It wasn't until Tuesday, January 17, that Ben Shattuck received the telegram from O'Neill reporting that his daughter had been found late in the day on Sunday—that she was alive but suffering from severe frostbite. Before he could leave Seward, Ben had to borrow money for the train tickets and other expenses of the journey. It took him two days to get from Seward to O'Neill by train, a distance of 135 miles. A week had now passed since Etta had gotten lost in the storm.

Ben Shattuck was familiar with gangrene from his three years of service in the Union Army during the Civil War. He knew, when he examined his daughter in the Murphys' farmhouse near Emmet, that the gangrene in her legs and feet was well advanced. There was no question of the girl walking.

It was Monday, January 23, before Ben and the Murphys felt Etta could be moved. The journey back to Seward was grueling. The Murphys arranged to have a sleigh filled with hay and they carried Etta out and tucked the hay around her, covering her over with a piece of canvas. The ride from the farm to the train depot in O'Neill was nine miles. Ben Shattuck walked at his daughter's side, despite the war wound in his right leg. The temperature that day never rose out of the single digits.

The train to Seward, scheduled to leave O'Neill early Tuesday morning, was eight hours late because of the weather. They thought Etta would be more comfortable if she could recline, so they arranged a kind of cot for her in the baggage car. The nightmarish trip took all night. Etta never once complained.

Poor as he was, Ben Shattuck spared no expense for his daughter's recovery. As soon as they got to Seward, three doctors and a nurse were called in. Drs. Reynolds, Potter, and Townsend saw at once that amputation was the only recourse. On Thursday, January 26, Etta lost both of her legs just below the knee.

The *Omaha Bee,* Tuesday evening, January 31, 1888: "The Shattuck Special Fund. Miss Etta Shattuck, the young school teacher who lost both limbs from the exposure in the recent storm will be incapacitated for any service by which she may derive a living. It is desired that $6,000 be raised. . . . This is to be known as the 'Shattuck Special Fund.'" The entire front page that day was devoted to blizzard relief. In light of the fact that Miss Shattuck's father was a war veteran who "suffered all the horrors of Andersonville prison" [actually, it was Belle Isle], every soldier was pressed "to send in his mite, be it ever so small, in appreciation of her heroic conduct and as a token of esteem for an old soldier and comrade." A new fund had been added—the Westphalen Monument Fund—to raise money for a headstone to be set over the graves of the Westphalen sisters.

Two days later, the *Bee* published a notice from Seward County Superintendent of Public Institutions George F. Burkett appealing for contributions to help Lena Woebbecke: "As yet nothing has been done . . . for the comfort and maintenance of Miss Woebbecke. She is the unfortunate girl who, on her return from school in district No. 71, northeast of Milford, became lost and remained out all night in a stubble field. . . . She is an orphan, eleven years old and is in very needy circumstances financially." The *Lincoln Journal* joined in the appeal. Children all over Nebraska handed over their pennies. Crusty typesetters at small-town newspapers coughed up a dollar.

Meanwhile, in the national press an unseemly brawl had broken out over the number of blizzard fatalities. On January 20, the *Nebraska City Press* devoted a story to a judge named A. F. Kinney, the agent for the Yankton Sioux tribe, who asserted that Dakota papers were deliberately underestimating and "covering up" the truth about the loss of life in the storm. When all the dead were counted, Kinney believed, the number would stand at almost one thousand in Dakota alone. The article went on to recount the details of Kinney's harrowing nine-day train journey through the storm—"The coal was running low. The passengers were crowded into one car trying to keep warm. Two babies perished. The men discarded all the outer garments they could spare and gave them to the ladies and children. Finding these not enough, they brought mailsacks from the post-car and wrapped the children up in them. While at one station in Bonhomme County {Dakota}, the Judge says, nineteen frozen bodies were brought into the depot in one day."

The *Chicago Tribune* picked up the story the next day, and it ran the day after that in the *New York Tribune.* The editor of the New York paper, however, seriously undermined Judge Kinney's claims by tacking on a vehement rebuttal from J. S. McLain, editor of the Minneapolis *Journal:* "This statement [of almost one thousand fatalities] is an evident absurdity. Future reports are rather likely to decrease than increase the list. . . . It is simply impossible that the news has been suppressed, for the *Journal's* reports have been verified by those of the morning papers. How a man who confesses to having been snowed up for the last nine days can estimate the loss of life with more accuracy than newspapers in constant telegraphic communication with the whole territory is incomprehensible."

What was at issue here was not just the accuracy of the death toll figures, but the truth about the climate of the prairie. A region that could slay a thousand innocent American citizens in the course of an afternoon did not look like a fit place for human habitation—quite the contrary—whereas if the figure stood at a mere couple of hundred, that could be written off as an unfortunate sacrifice on the path to progress. In essence it was an argument over image and reputation: prairie public relations.

Dakota and Nebraska papers rose as one to defend their land and its climate. Editorials insisted that even with the blizzard fatalities their territory remained far healthier and enjoyed a far lower death rate than the fetid cities of the East Coast or the malarial swamps of the South. "Let people who are worrying about loss of life in Dakota remember this," demanded the *Sioux Falls Argus Leader.* "In Dakota a man occasionally gets blown away or frozen to death. But his teeth never chatter for years because of ague, his form never withers under miasmatic troubles and seldom does his face whiten with pallid consumption. . . . Dakota is, once in years, the scene of some distressing calamity such as saddened her homes last week. But the lingering sickness and epidemic disease so common in other places are strangers here." There were catalogs of Atlantic storms, Pacific earthquakes, and lake region "catastrophes" in the *Aberdeen Daily Republican.* The *Daily Huronite* refused to grieve over the losses, but instead greeted the blizzard in a "spirit of rejoicing." "And why? Because the deep snow means good harvest in Dakota: the drought deamon [*sic*] flies before the breath of the snow blizzard."

The *Mitchell* (Dakota Territory) *Capital* growled that reports in the Eastern papers of the death and destruction inflicted by the blizzard were not only grossly exaggerated, but deliberate attempts to "disadvantage . . . the entire territory of Dakota." The message was clear: Devious editors were using the blizzard to blacken the good name of the prairie in order to scare away prospective settlers or divert them to their own regions. The *Capital* clinched its defense of Dakota with hard data. Though there was no denying that the storm had been deadly, the paper's editor calculated that when the number of fatalities was averaged over the entire population of the territory, there was but a single life lost for every three thousand residents.

Just about the same ratio applied to New York City on the morning of September 11, 2001.

The heroine story took on a life of its own. Papers were now advertising a large photo of Minnie Freeman posing by her sod schoolhouse with the precious children she had saved—a dollar each, eight dollars for a dozen. The Roll of Honor expanded daily. The *Bee*'s lead story on February 4 opened with a detailed description of the condition of Lois Royce's feet—"a great piece of frozen flesh has sloughed off from one side of the heel"—prior to their amputation.

But the greatest coverage by far—and the largest share of the money raised—went to Etta Shattuck. Anything written or said by anyone remotely connected with her was rushed into print. "I know the family and I have known Miss Etta," George W. Morey, a minister living in Wahoo, Nebraska, confided to the *Bee* on January 30. "I have been their pastor for two years and was personally acquainted with them and their circumstances before they went to Holt county, Nebraska two years ago. Mr. Shattuck is not only a veteran soldier but a worthy upright honorable Christian gentleman having the respect of all who know him. I knew Miss Shattuck while in Seward as a brave young woman struggling to prepare for the work of teaching with aught but their own labor to aid her. Who can tell the suffering of those 'seventy-eight hours!'"

Two days later the Shattuck family's current pastor, the Reverend J. H. Presson of the Seward Methodist Episcopal Church, sent in his own firsthand account of a recent conversation with Etta. "The religion of Christ sustained me in this affliction," Etta had told him solemnly. "I have suffered but very little pain, for which I thank God." The Reverend Presson was convinced that Etta's life "was spared that she might show us how victorious a christian [*sic*] can be."

Etta's doctors, however, remained cautious. Though she seemed stronger in the first few days after her operation, it was becoming clear that the damage from the frostbite had progressed farther than Drs. Reynolds and Potter had originally thought. A second round of amputations now looked unavoidable. In addition, a wound had opened up in the part of her back that had been in contact with the hay during her three-day imprisonment. The skin and the flesh beneath the skin were falling away, leaving a two-inch-deep cavity. Healing these kinds of wounds was extremely difficult.

The Reverend Presson again reported that Etta was remarkably brave and cheerful. But on Thursday, February 2, she took a turn for the worse, growing suddenly feeble and listless. Drs. Reynolds and Potter had feared that this might happen. Like all nineteenth-century doctors, they had seen the pattern before: A patient would survive surgery, show promising signs of recovery, but then swiftly go into a decline. In most cases the cause was sepsis—an overwhelming bacterial infection of the bloodstream contracted during surgery or from the unsanitary condition of hospitals or homes. It's also conceivable that Etta had come down with pneumonia as a consequence of her prolonged exposure to cold in a weakened condition. In both cases, the doctors were powerless to help her. Without antibiotics there was little they could do but keep her comfortable and prepare her for the end.

Etta told her pastor that she was perfectly satisfied no matter what her fate might be. Her mind remained clear and her faith unshaken. She had no word of regret and gave no sign of fear.

Early on the morning of Monday, February 6, Reynolds and Potter were summoned to the Shattucks' house. The doctors knew at a glance that death was imminent. Etta's blood pressure had dropped precipitously, her breathing was light and shallow and her pulse rapid. Delirium clouded her few waking moments.

Ben and Sarah Shattuck gathered at their daughter's bedside with their four surviving children. Etta died peacefully at 9 A.M., two months and two days shy of her twentieth birthday.

CHAPTER TWELVE

Aftermath

Drexel & Maul, Omaha's premier undertaking firm, selected a casket—"the most expensive casket they have," according to The *Bee*—and sent it free of charge to Seward on the day Etta Shattuck died. The funeral was held the following day, Tuesday, February 7, at 2 P.M. Seward's Methodist church was packed to overflowing hours before the procession arrived from the Shattuck house. Members of the local Grand Army of the Republic post—the Seward veterans of the Civil War—led the cortege. Six young men bore Etta's casket, which had been draped with a cloth of white velvet brocade and fixed with a plate that read "At Rest." Members of the Seward press had contributed a floral display of a two-foot-high arch from which hung the gates of heaven, left ajar to receive Etta's soul. The scent of lilies pervaded the church.

One should not speak of Miss Shattuck as dead, the Reverend Presson insisted in his sermon, for her heroism and religious spirit made her life appear more like the morning twilight than the shadows of evening. While the congregation wept, the minister spoke at length of Etta's faithful discharge of her duties as a daughter, a teacher, a member of the church, and the sole support of her family. After the sermon the mourners filed by the open white casket.

It had thawed the first week of February, but that day the temperatures were back in the single digits and the ground was frozen hard. It was a long, cold mile from the Methodist church to Seward's North Cemetery. The funeral procession moved in a dark line between the white fields. Etta Shattuck was laid to rest among the graves of Seward's Civil War veterans.

None of the funerals of the other storm victims was remotely like this. Parents who lost children to the storm summoned carpenters or neighbors to nail together coffins of whatever lumber they had on hand. It was the custom among some of the immigrant families to blacken the coffins with burnt cork. "On two chairs stands the little black coffin," wrote a Norwegian mother of a child's funeral in Dakota. "With awe we had watched the carpenter, working with candle light, the night before, making this box. We had seen the little form laid into it and the lid nailed down." The child's father and the carpenter sang a hymn, and then they carried out the coffin and put in on the back of a sleigh. "The father bade the wife good-by, threw the sleigh rope over his shoulder and, followed by the carpenter, started off on the weary trudge to the cemetery. That was a pioneer funeral."

It was worse than that for the three Schweizer families who had lost sons in Rosefield Township. Even after the bodies of their boys had thawed, rigor mortis kept them in the contorted positions in which they had died. The parents had to struggle to get their sons into coffins. The five boys were buried in a single grave in the cemetery next to the Salem Mennonite Church, three miles from where they died.

It was so cold in Brookings County, Dakota, that the Tisland family could not dig a proper grave for the body of their father, Ole. His son, Ole Jr., tried day after day to get his team to the lumberyard at Volga to buy a coffin, but finally gave up. Using scrap lumber and boards pried from a straw shed, Ole Jr. managed to make the box with the help of the neighbors and lined it with a quilt. When they finally got the grave dug, it snowed, so they had to dig the snow out of the hole before they laid the old man to rest. On the way to the cemetery the sleigh tipped over as the horses were hauling the coffin through the snow. Finally the men set Ole Tisland's coffin on a toboggan and dragged him to his grave. Most of the women turned back on account of the cold, but Ole's wife and his daughters followed behind the toboggan, walking in the footsteps of the men. One of the neighbors said the Lord's Prayer in Norwegian by the open grave.

It was spring before the Lutheran pastor came and offered a proper Christian burial service at Ole Tisland's graveside.

George Burkett, the superintendent of public institutions for Seward County, made it his personal mission to save the life of Lena Woebbecke. Thanks to Burkett's efforts, the *Bee* ran the story of how the German girl had spent the night of the blizzard alone on the open prairie and began raising money on her behalf. Dr. G. W. Brandon was summoned from Milford to attend to the child. He reported that the Woebbecke family was destitute and Lena was suffering terribly from frostbite. Gangrene was well advanced in her feet and Dr. Brandon amputated her right foot above the ankle at once. The left foot, he believed, could be saved. Lena survived the surgery but in the days afterward complained of excruciating headaches. A local paper reported that the child was "greatly isolated where she is at present, the roads in that vicinity being almost impassable. She ought to have better clothing and better nursing."

Eventually Burkett secured a court order to be appointed Lena's guardian and trustee of the nearly four thousand dollars that had been raised for her—a sizable sum at a time when a hired man on a farm earned two dollars a day, a hotel clerk three dollars a day, and a housekeeper counted herself lucky to get two dollars, fifty cents a week. Lena was moved from the Woebbecke house and placed in the care of a German woman in Milford. In time she was fitted with a wooden foot and began walking again.

❧❧

By the end of February, the Heroine Fund had reached $11,267.56. The *Bee* reported that the $3,752.01 that had been raised for Miss Shattuck would now go to her bereaved family. Minnie Freeman, in addition to her share of the Heroine Fund, received a gold watch from a prominent San Francisco jeweler: The watch chain was fashioned to look like a rope, a tribute to the teacher's foresight and bravery in roping her students together and leading them through the storm (though at least one of these students denied that this had ever happened). Lyon & Healy, a Chicago music publisher, printed a song about Minnie's heroic deeds—"Song of the Great Blizzard 1888, Thirteen Were Saved or Nebraska's Fearless Maid," dedicated to Miss Minnie Freeman, schoolteacher and heroine, "Whose Pluck and good Judgment Exhibited during the Recent fearful Blizzard in the Myra [*sic*] Valley District, Nebraska, Saved the Lives of thirteen helpless little Children." Eighty some odd men wrote to ask her hand in marriage. Minnie became a national celebrity, her image in wax displayed throughout the country, her story retold and elaborated endlessly. The legend endured for years. Well into the twentieth century, whenever Nebraska was mentioned in other parts of the country, it was Minnie Freeman and the blizzard that sprang to people's minds. As one prominent Nebraskan wrote in his pioneer memoir, "Twenty years after the storm I visited St. Petersburg, Florida, and heard the Minnie Freeman story with all the horrors of an Arctic storm accompaniment being discussed among those at the table where I ate my first meal in that town."

"We had been calling every storm a blizzard," the Reverend S. F. Huntley wrote a friend in New York, "but then [after January 12] decided that we had never had a blizzard before and never wanted one again." The three Huntley children recovered from the touch of frostbite they had suffered during the storm, as did Fred Weeks. By all rights Fred should have been lionized as a hero of the storm, but for some reason local papers failed to take up his story. However, a fund was set up to raise money for Addie Knieriem after her foot was amputated. Through the efforts of A. M. Mathias, the doctor who had performed the amputation, a philanthropist in Brooklyn, New York, learned of Addie's plight and agreed to set up a yearly annuity of six hundred dollars to aid the child. Like Lena, Addie eventually was able to walk unassisted on a wooden foot.

Some of the money that was raised for the Westphalen sisters went to purchase the soaring marble obelisk that marks their graves in the cemetery of St. John's Ridgeley Lutheran Church in the hamlet of Ridgeley, Nebraska. Three couplets inscribed in the stone commemorate the brief lives of Eda and Matilda Westphalen:

> *How soon alas our brightest prospects fail*
> *As autumn leaves before the driving gale;*
> *Meteors an instant glittering through the sky*

Like them they fall but not like them they die:
In cloudless glory they shall ever bloom
New life inhale immortal from the tomb.

But even in the grave weather was not done with the sisters. In the summer of 1890, a windstorm blew their monument over and broke the obelisk in half. "All who contributed will regret to hear of the circumstances," reported the local paper. A new collection must have been taken up because the obelisk stands flawless today.

❧ ❧

"There was a cruel aftermath to the blizzard," wrote a survivor, "funerals, surgical operations, cripples, fingers with first joints gone, ears without rims, and some like poor Will Moss, who spent the night on the prairie in the shelter of his cutter, and supposed that he had escaped without damage, afterward died of diseases caused by the exposure."

The precise number of the dead was never determined. Estimates published over the years in state histories and local newspapers have ranged from 250 to 500. The southern and eastern part of Dakota Territory suffered the majority of the casualties. Undoubtedly many deaths were never reported from remote outlying districts. Scores died in the weeks after the storm of pneumonia and infections contracted during amputations. For years afterward, at gatherings of any size in Dakota or Nebraska, there would always be people walking on wooden legs or holding fingerless hands behind their backs or hiding missing ears under hats—victims of the blizzard.

❧ ❧

Sergeant Glenn wrote that the widespread loss of livestock "will operate seriously against the losers, as they are principally men struggling to make a home in a new country. Care of and anxiety for livestock was the principal cause which took men away from the immediate vicinity of their homes."

Huron and the surrounding country was especially hard hit, and in the days after the storm angry farmers demanded to know why the Signal Corps had failed to issue coordinated storm warnings far enough in advance to do them any good. "The service has been criticized in several quarters in the Territory for not disseminating a specific warning from some central point, like Chicago or Saint Paul," Glenn wrote in his storm report. "So far as he has had the opportunity, the observer has explained that the Service can not do more than it does in consequence of contracted means in the way of appropriations." This has a decidedly contemporary ring to it. In fact, Glenn admitted that the "first definite information that a storm of such magnitude was approaching" came not from the Signal Corps, but over the railroad telegraph wires. Glenn hinted that the heavily centralized bureaucratic structure of the Corps itself was at fault for the forecast failure since observers like himself were unable to make use of their knowledge of local weather conditions. Back in the autumn, when Lieutenant Walshe had inspected the Huron station, Glenn requested permission "to have an opinion about the weather"—in other words, to issue his own local forecasts. He now speculated that had such authority been granted, instead of expressly denied in his orders from General Greely, he might have been able to warn the people of Huron and its vicinity of the coming storm.

Could Sergeant Glenn have saved the life of Robert Chambers, the Huron farmer who sacrificed himself so that his son might live, had his hands not been tied by military regulations? Probably not, given the primitive communication system at the time. Still, it was maddening to Glenn that in the face of meteorological disaster Army regulations rendered him essentially powerless.

Greely's first impulse was to downplay the severity of the storm. The *Monthly Weather Review* for January 1888, published by the Signal Office in Washington under Greely's direction, conceded that it was "a violent storm in which many lives were lost and large numbers of cattle perished," but then went on to argue that "the loss of life as given in newspaper accounts has been doubtless exaggerated, but was evidently greater than in any previous storm, owing to the extensive settlement of the country in the last few years." Only later, when he sat down to write his book *American Weather,* did the general acknowledge that the storm was "the most disastrous blizzard ever known in Montana, Dakota, Minnesota, Kansas, and Texas. The change in the direction of the wind and the fall in the temperature was more sudden than usual. . . . High winds ranging from thirty to fifty miles per hour occurred, with falling and drifting snow, which, in addition to the great loss of human life, caused the destruction of herds of cattle and an enormous amount of suffering to entire communities."

Today a "surprise" storm that killed over two hundred people would instigate a fierce outcry in the press, vigorous official hand-wringing, and a flood of reports by every government agency remotely involved, starting with the National Weather Service. But in the Gilded Age, blame for the suffering attendant on an act of God was left unassigned. Hardly anyone believed that government agencies had either the expertise or the obligation to forestall disaster, and that attitude suited the government just fine. Though the Southern newspapers muttered a bit about the Signal Corps' failure to provide sufficient warning of the cold wave, the Northern press refrained from finger-pointing. Heroines were called for, not culprits. Making too big a deal about the storm would only scare away prospective settlers or send them fleeing to balmier climes. Best to put the whole incident behind them and get back to work.

In any case, the odds of such a violent storm striking again were extremely low. As Greely himself wrote on January 26, quelling the fears of a citizen who had inquired whether blizzards were becoming worse, "There is no reason to believe that there is any secular change in their frequency or intensity."

❦ ❦

The only head that rolled in the wake of the storm belonged to First Lieutenant Thomas Mayhew Woodruff.

In the few months he had been forecasting the weather from Saint Paul, Woodruff had made several bitter and influential enemies—notably Professor William Payne and Thomas Cochran Jr. of the Saint Paul Chamber of Commerce—who were determined to make him pay for encroaching on their turf. And in the course of that winter and spring, they succeeded. Greely, characteristically, made the situation worse by barking orders from Washington, privately offering support to both sides simultaneously and ultimately caving to superior muscle and money. Had Woodruff been a better forecaster, the general might have done more for him. But since Woodruff's "percentages of verifications of indications" (in other words, how often he got the forecast right) were among the lowest in the Corps that winter, Greely had little incentive to step in. When it came down to it, Woodruff was outmaneuvered and Greely stood by and watched him swing. Neither meteorological competence nor military honor nor professional loyalty was really the issue: In the end, it was just politics as usual.

On February 14, a week after Etta Shattuck's funeral, Greely pressed Woodruff to file a full report on his wrangle with Professor Payne and the Meteorological Committee of the Saint Paul Chamber of Commerce. Woodruff's report, duly sent to Washington on February 17, rehearsed the whole sorry business: his turf war with Payne over the status and standing of the Saint Paul indications branch office vis-à-vis the Minnesota State Weather Service; the embezzlement charges that were being pressed against Cochran; the sloppy accounting of the Chamber of Commerce and its murky relationship with the state weather service; the "begging circulars" sent out by Cochran to solicit contributions for the support of the State Weather Service from local businesses; the resentment that railroad officials harbored against Payne, etc.

Meanwhile, Payne retaliated by complaining of Woodruff's failure to open additional weather stations along the railroad lines, which he had been expressly ordered to do when Greely posted him to Saint Paul. Woodruff defended his actions and questioned Payne's judgment. Payne counterattacked. By winter's end, a steady stream of accusation was flowing from Saint Paul to Washington. Eventually things got so ugly that Greely ordered Woodruff to complete a questionnaire about the nature of his conflict with Payne and Cochran. When he sent his answers, Woodruff insisted to Greely most forcefully that all of his communications on this matter be kept strictly confidential. "I certainly have no objections to having either Prof. Payne or Mr. Cochran know my opinions of them either personally or officially," Woodruff wrote, "but I should most certainly strongly protest against having my report[s] . . . submitted to either or both of them."

Greely evidently disregarded this plea. In mid-April, the general instructed Woodruff to close the Saint Paul indications office and return to Washington since, as he put it, "the season of cold waves [has] substantially passed" and therefore an on-site forecaster was no longer needed. At the same time, Greely, through an underling, made Woodruff's confidential reports and communiqués available to Professor Payne, who in turn forwarded them to the Saint Paul Chamber of Commerce. On May 1, while Woodruff was packing up his office files and seeing to the storage of the furniture, the Saint Paul Chamber of Commerce met to consider the charges that Woodruff had lodged against Cochran, Payne, and the state weather service. The upshot of this meeting was that Cochran and another Chamber member were delegated to travel to Washington to see Greely and discuss the matter in person. No record survives of what happened at this meeting, but the consequences speak for themselves. Immediately afterward, the secretary of war wrote to Greely to inquire whether Woodruff's services are of "such a nature that any injury to the service will result from his being ordered to join his regiment." On May 8, Greely directed an assistant to write a memo to Woodruff (who was then on an eastbound train near Chicago) demanding that he immediately "investigate and make written report regarding the causes that delayed, retarded, or prevented the expected advent at stations of cold waves and other weather conditions predicted by yourself." This report was to "embody . . . such special facts as may have substantially come to your notice, which at the time escaped the attention of the officer making predictions, and which if noted might have materially aided in indicating approaching changes."

It was surely not a coincidence that the first time Greely broached the subject of the late arrival of storm warnings came immediately after his meeting with Cochran. Not a word of criticism of Woodruff's forecasting skills or communication network all winter long, not a whisper of complaint over "delayed or retarded" warnings after scores of children perished in subzero temperatures on the night of January 12, no memo demanding why the order to hoist cold wave flags at Signal Corps stations in Dakota and Nebraska arrived only minutes before "the most disastrous blizzard ever known," or in some cases afterward. Yet now that Woodruff had closed the indications office for the season (actually for good, as it turned out), an urgent message arrived telling him to report on his meteorological failings as soon as possible. Payne had been savaging Woodruff all winter for not seizing every means at his disposal for disseminating forecasts and warnings in a timely fashion, no matter what the cost to the government—but it wasn't until Cochran himself turned up in Washington to do some political arm-twisting that Greely took up the cry. The "causes that delayed, retarded, or prevented" the timely arrival of forecasts and the "special facts" that "might have materially aided in indicating approaching changes" were just excuses: At the time of the blizzard, Cochran and Payne had shown precious little interest in the effect of delayed warnings on the farmers and children and teachers of the prairie region. They wanted Woodruff bounced out of the service for petty personal reasons and they were handing Greely the ammunition he needed to do their dirty work.

Woodruff's fall was cushioned by a new commission that conveniently came his way. On May 23, the day after Greely reported to the secretary of war that "no injury would result from [Lieutenant Woodruff's] being ordered to rejoin his regiment," Woodruff informed the adjutant general that he had been offered the position of aide-de-camp by Brigadier General Thomas H. Ruger and that he wished to take the offer, which would require him to leave the Signal Corps and rejoin his regiment. Ruger, a fellow graduate of West Point and a longtime career officer who had commanded a division at Gettysburg and battled draft rioters in New York during the Civil War, had been headquartered at Saint Paul as commander of the Department of Dakota during the time that Woodruff was stationed there. The two men no doubt became friendly in the course of the winter and it's possible that General Ruger, a cultivated, quiet-spoken, gentlemanly officer like Woodruff himself, extended an open invitation to join his staff if Woodruff ever found himself at loose ends.

Even though Woodruff was swiftly able to arrange an attractive fall-back position, the fact remains that he was unhappy about leaving the Signal Corps and angry at the forces that had been working behind the scenes to get him out. He may have been angry at Greely as well for failing or refusing to help. Certainly, relations between Greely and Woodruff grew noticeably stiff and formal. On the day after Woodruff left the Signal Corps after five years of service, Greely, through an underling, wrote to express his tepid gratitude for Woodruff's "hearty and loyal support" and to wish him "pleasant and agreeable duty in other paths of military life, where he [the general] feels certain that the soldierly qualities shown by you in connection with this service will ensure you that promotion and reward denied you in common with other line officers whose work has so materially contributed to the great success of the weather bureau." Woodruff's primary concern now was with the contents of his military file, which would have a bearing on future promotions, and Greely was barely cooperative in this matter. The general grudgingly agreed to add to Woodruff's file a "commendatory letter" signed by some of the most prominent citizens of Saint Paul attesting to the "able manner" in which Woodruff had performed his duty "in the interest of the Northwest." But he retracted an earlier promise to supply Woodruff with copies of other documents in his file, stating curtly that papers he had submitted to the War Department "preclude any report on affairs at Saint Paul."

Greely in short had washed his hands of Woodruff and his future in the military. As far as the general was concerned, the case was closed. Woodruff would never again forecast weather.

❧ ❧

Meanwhile, exactly two months after the January 12 storm, the Signal Corps botched the forecast of another, even deadlier blizzard. On the morning of Monday, March 12, heavy rain changed to heavy snow over New York City and temperatures began to plummet. In the course of the day, twenty-one inches of snow piled up in the city. Winds gusted to 50 miles an hour, throwing up twenty-foot-high drifts, snapping telegraph poles, and snarling the live telegraph and telephone wires that the poles carried. Elevated trains and carriages were stopped in their tracks. By afternoon the nation's largest and most modern metropolis was at a standstill: No vehicles moved in the streets, no communications got in or out of the city, no power flowed to buildings. Outside the city, in southern Connecticut and upstate New York, nearly four feet of snow accumulated. Tens of thousands were stranded. An estimated four hundred people died, two hundred of them in New York City. "For the first time in their lives [New Yorkers] knew what a western blizzard was," reported the *New York Times*. "The city was left to run itself," wrote the *Tribune;* "chaos reigned, and the proud boastful metropolis was reduced to the condition of a primitive settlement."

The Signal Corps station on the top floor of New York's nine-story Equitable Building—one of three branch offices then authorized to issue local indications, along with the Saint Paul and San Francisco offices—was closed from midnight Saturday until 5 P.M. on Sundays. The last set of indications issued Saturday night called for rain Sunday followed on Monday by "colder fresh to brisk westerly winds, fair weather." By the time the New York station chief, Elias Dunn, and his staff reported for duty on Sunday evening, the telegraph connection to the Signal Office in Washington was dead. So no word arrived in New York that a wild nor'easter was moving up the coast, that twenty-seven of the forty ships anchored at the Delaware Breakwater had been seriously damaged or destroyed, and that scores of other ships were sinking and running aground from Chesapeake Bay to Nantucket.

One of the men on duty at the New York Signal Corps office at the time of the storm was the redoubtable German immigrant Sergeant Francis Long, a survivor of Greely's Arctic expedition. Greely, who always looked out for his own, had secured Long the berth in New York. The New York press was not kind about the Corps' failure to predict the storm, but Greely saw to it that Long kept his job through the maelstrom.

Two positive changes came about as a result of the East Coast blizzard: Signal Corps weather stations began to remain open all day on Sundays and eventually the unsightly webs of the city's telegraph and telephone wires were taken off the streets and moved underground.

This colossal forecasting error coming hard on the heels of the January 12 blizzard was not good publicity for Greely or the Signal Corps. The weather service had become a national embarrassment. The Corps had demonstrated conclusively that it was all but helpless in the face of meteorological disaster. But it had even less utility as a military organization. Army brass raged that weather now consumed so much of the division's time and budget that the armed forces would be left without an effective communications network in time of war. Scientists wondered why Signal Corps fixtures like H.H.C. Dunwoody, an oily first lieutenant who had spent years ingratiating himself to various chief signal officers, were given free rein to gather weather proverbs (Dunwoody, with Corps sponsorship, published an entire volume of "popular weather sayings" in 1883) while serious research was ignored, neglected, or starved of funds. There had been repeated calls in Congress and by the secretary of war to transfer the weather service from the Army to the Department of Agriculture, but so far the Signal Corps lobby had succeeded in blocking the measure. Greely himself, though well aware of the flaws of the present system, waffled on the issue of transfer.

Finally, in December 1889, President Benjamin Harrison formally recommended that the responsibility for weather forecasting and data gathering be moved from the Army Signal Corps to the Department of Agriculture, and this time both houses of Congress, having repeatedly blocked the measure in the past, swiftly agreed. What motivated the congressional change of heart was the conviction that the transfer would make farmers happy and perhaps inspire them to put more trust in the weather service. Scant mention was made of public safety.

Congress passed the bill with little discussion in April 1890, and the president signed it into law on October 1, 1890. On July 1,

1891, General Greely ceased to have jurisdiction over the weather of the United States and the ill-fated Professor Mark W. Harrington became the first civilian chief of the new United States Weather Bureau.

❦ ❦

Years after the blizzard, Mary Matilda Sisson of Dakota's Douglas County told her daughter that once the snow melted that spring they found dead horses and cattle and the bodies of several men and schoolchildren on the prairie near their homestead. But despite the heartbreak, Mary remembered that "Never was spring more beautiful. The birds came back, the flowers bloomed and the grouse and prairie chickens boomed and strutted on the knoll northwest of our house."

That lovely soft spring Johann Kaufmann Sr. was out mowing the tall luxuriant grass that had grown in his fields. Since his sons died, Johann had had trouble concentrating and he suffered from kidney trouble. Distracted by his pain, he was not paying attention as he worked the horse-drawn mowing machine with its treacherous five-foot-long blades back and forth through the tall grass. He accidentally ran the mower over a young boy who was visiting the farm and cut off his leg. There were bitter words and recriminations from the boy's family. It was another terrible blow for the Kaufmanns. Two years later, grief-stricken and bereft, Johann Kaufmann died.

Johann's wife, Anna, a widow who had lost her first six children, remarried and moved to Kansas. But it was not a happy match. The second husband was unkind and her four surviving children—Julius, Jonathan, Emma, and Anna—did not get along with their stepfather. Julius died in Kansas of a ruptured appendix at the age of twenty.

Jonathan married to get away from the stepfather and returned to his childhood home in Dakota's Rosefield Township to raise a family. His two daughters, Gladys and Anna, were born there, Gladys in 1912 and Anna in 1917, and they still live near the old homestead in the town of Freeman, South Dakota. Jonathan, three and a half years old when his three brothers died, rarely spoke to his children about the blizzard. Gladys and Anna say that the only thing they remember their father telling them about that day was the sight of the three frozen bodies lying next to the stove—and the sound of his mother's laughter.

In the autumn of 1888, George Burkett moved his ward, Lena Woebbecke, to Lincoln and enrolled her at the C Street School. Burkett reported that the child's English was still poor, but he expected her to make good progress now that she was attending a grade school instead of a one-room country school where children of all ages were mixed together. Lena had learned to walk quite well on her wooden foot.

Burkett invested $3,750 of the money raised for Lena in real estate secured by first mortgages with a handsome return of 8 percent. He was confident that this would generate more than enough income to support and educate the girl.

For the next six years, Lena lived in Lincoln under Burkett's supervision. She graduated from the public school and then attended Union College, run by the Seventh Day Adventists. When she turned seventeen, Burkett made over to her the $4,939.46 in cash and notes that had accrued from the investment of her fund. In gratitude, Lena and her family gave him a beautiful rocking chair. The family planned to invest Lena's money in a farm near Milford.

The trail of Lena Woebbecke's life becomes faint in her final years. In 1901, when she was twenty-four years old, she married a local man named George Schopp, a German by the sound of his name and most likely a farmer. She died less than two years later at the age of twenty-five—whether from disease or accident or some lingering complication of her amputation or in childbirth, as was all too common in those days, we'll never know. Lena was laid to rest in her wedding dress in the graveyard of the Immanuel Lutheran Church near the country crossroads called Ruby. If there ever was a town of Ruby, it has disappeared, as has the Immanuel Lutheran Church. The church cemetery, however, remains—a fenced patch of rough grass studded with headstones between two farmhouses not far from the interstate. A tiny island of the dead in the sea of Nebraska agriculture.

Lena's mother, Wilhelmine Dorgeloh, died a few days after her daughter at the age of fifty-two. She was buried beside the grave of the child she had abandoned sixteen years earlier. Their matching granite headstones—inscribed in German and decorated with sprays of chiseled leaves—are by far the finest in the churchyard, a final legacy of the heroine fund. *"Wahrlich, wahrlich ich sage euch: so jemand mein Wort wird halten, der wird den Tod nicht sehen ewiglich,"* reads the inscription on Lena's headstone. "Verily, verily, I say unto you, If a man keep my saying, he shall never see death."

As for the Woebbecke family that took Lena in during the summer of 1887, their descendants are still living and farming the same hilly acres south of Seward. Lawrence Woebbecke, the grandson of Wilhelm, who carried Lena up through the ravine on the morning of January 13, grows wheat and corn and soybeans on the family farm. He does well enough that he was able to travel to Germany a few years ago and visit the village of Herkensen outside of Hameln that his grandfather left in 1878. A married son lives nearby and helps with the farmwork, and there are grandkids, too—so it looks as if Woebbeckes will be on this land for some years to come.

※ ※

Walter Allen never forgot that he owed his life to the scrappiness and determination of his brother Will. The boys remained close for the rest of their lives.

Eighteen at the time of the blizzard, Will Allen already had six years of newspaper printing experience under his belt in Groton and he soon sought out the larger challenges of Aberdeen, recently forsaken by the hapless L. Frank Baum. Unlike Baum, however, Will Allen made a notable success of his journalistic enterprises in Aberdeen. Eventually he went to work for the *Dakota Farmer,* the region's premier agricultural publication, and over the years he rose through the ranks to become managing editor, editor in chief, and publisher. In 1933, on the strength of his reputation for integrity and hard work and his wide circle of business acquaintances, Will Allen secured the nomination of the Republican Party in the South Dakota governor's race. He ran as a progressive Republican on a platform of operating government with the efficiency of a business, restoring the state's property tax, and bringing nonresident landowners onto the tax rolls. The Democratic candidate, Thomas Berry, defeated him resoundingly. Allen died six years later at the age of sixty-nine.

Will's younger brother, Walter, got himself into several more scrapes after the blizzard, but he managed to survive his childhood, graduate from the University of Minnesota, and land a couple of jobs with railroads before he joined his older brother on the staff of the *Dakota Farmer* in 1910. Walter remained with the publication for the next fifty years, retiring as its director in 1960. He died in 1973 at the age of ninety-three.

Walter's daughter, Barbara Wegner, still lives in Groton, almost directly across the street from the long-demolished schoolhouse where her father went to fetch his precious perfume bottle on the afternoon of January 12, 1888.

Lieutenant Woodruff was six months shy of his fortieth birthday when he left the Signal Corps on June 1, 1888, a vigorous if belea-guered infantry officer with a wife and young daughter and a mili-

tary career that had been stalled since 1879 at the rank of first lieutenant. Eleven years later he was dead.

For some reason Woodruff did not take up his post as General
Ruger's aide-de-camp immediately but instead returned to his regi-
ment for fifteen months of frontier service at Fort Bliss near El
Paso, in the extreme western tip of Texas. When he joined Ruger at
Saint Paul in August 1889, Woodruff installed himself in the
swanky new Aberdeen Hotel up on the bluff near the mansions of
Summit Avenue. "Bathroom with each apartment," boasted the
Aberdeen's ads. "A high-class patronage solicited. Rates $3.50 to
$6.00 per day." There is no record of whether Woodruff crossed
paths with Professor Payne or Mr. Cochran during his second resi-
dence in Saint Paul, but it seems unlikely.

Two years later, the long-awaited promotion to captain finally
came through. Woodruff left Ruger's staff and returned to his regiment, now stationed in Florida.

Shortly after President McKinley declared war on Spain on
April, 25, 1898, Woodruff was appointed Inspector General of Vol-
unteers. He served with the obese, bumbling General William
Shafter during the Santiago campaign, the decisive land and sea
battle for Cuba's second-largest city, and was later on the staff of
Major General John C. Bates. The Spanish-American War was brief
by nineteenth-century standards—Cuba fell to U.S. forces by July
and a cease-fire was declared on August 12—and relatively inex-
pensive. The US government laid out about $250 million on com-
bat and ended up with Cuba, Puerto Rico and Guam; for another
$20 million Spain handed over the Philippines. Of the three thou-
sand Americans who lost their lives in the war, 90 percent died of
infectious tropical diseases, especially malaria and yellow fever.

One of the victims of disease was Captain Woodruff. He was still
in Cuba the summer after the war when he contracted yellow fever
near Santiago. He died at the age of fifty on July 11, 1899. A loyal
soldier, Woodruff had devoted twenty-eight years of his life to serv-

ing the dreams of his young ambitious country. On March 5, 1900, his remains were taken to Arlington National Cemetery and laid to rest in the presence of his widow, his daughter, family and friends.

❧ ❧

"I have seen the Dread of Dakota. A genuine blizzard and am now ready to leave anytime, that we can sell," pioneer wife Sadie Shaw wrote to relatives back east from her Dakota homestead in Douglas County. *"Oh, it was terrible.* I have often read about Blizzards but they have to be *seen* to be fully *realized."*

The Shaws did not sell, nor did Abi Huntley prevail upon her husband to take her and their traumatized children back to New York State. There were many such threats and much misgiving after the blizzard of 1888, but few families left—at least not right away. The weather finally moderated. Summer came and the prairie turned hot and dry. Day after day the sun sucked the moisture out of the black soil of the prairie. Grieving families got on with their lives, prayed for rain, had more children.

The blizzard of January 12, 1888, did not put an end to the great white endeavor of settling and taming the prairie, but it did mark a turning point, a change of mood and direction. The Dakota boom had ended. Immigration to the prairie frontier slowed to a trickle in the last years of the 1880s. A time of reckoning and taking stock had set in. A new mood of caution, suspicion, and bitterness took hold. "Good bye, Lord, I am going west," Arthur Towne remembered the church deacon shouting as Dakota-bound families streamed out of their Vermont village in 1881. By the close of the decade the joy was gone and the Townes were exhausted. "It did seem as if the whole James River valley was just a dumping ground for blasted hopes," Towne's mother told him wearily. "The holiday spirit of eight years before had entirely vanished," wrote Hamlin Garland of the sullen mood of the decade's end. "The stress of misfortune had not only destroyed hope, it had brought out the evil

side of many men. Dissension had grown common. Two of my father's neighbors had gone insane over the failure of their crops. . . .
[S]omething gray had settled down over the plain. Graveyards,
jails, asylums, all the accompaniments of civilization, were now
quite firmly established. . . . No green thing was in sight, and no
shade offered save that made by the little cabin. On every side
stretched scanty yellowing fields of grain, and from every worn
road, dust rose like smoke from crevices."

The truth was beginning to sink in: The sudden storms, the violent swings from one meteorological extreme to another, the
droughts and torrents and killer blizzards were not freak occur-
rences but facts of life on the prairie. This was not a garden. Rain
did not follow the plow. Laying a perfect grid of mile-sided squares
on the grassland did not suppress the chaos of the elements. The
settlers had to face the facts. Living here and making a living off
this land was never going to be easy.

Weather that takes lives and destroys hopes presents a moral
quandary. Call it an act of God or a natural disaster, somebody or
something made this storm happen. But what? Who was to blame
for the deaths of the Kaufmann brothers? For the surge of frigid air
that killed John Jensen's wife, Nickoline, and their little daughter,
Alvilda? For the fact that after January 12 Addie Knieriem never
again walked on the feet she was born with, never ran, woke up every
morning to the sight of the scarred stump below her ankle? Was it
the fault of the railroads and the United States government for col-
luding to lure pioneers to country too wild and dangerous to support
secure settlements? Was Lieutenant Woodruff guilty for failing to see
the storm coming sooner or for not striving harder to get the word
out? Were the immigrant parents themselves to blame for uprooting
their families from the relatively safe enclaves of the Ukraine, Vermont, Prussia, and Norway and exposing them to the brutal cold
fronts and lows that sweep down off the Canadian Rockies?

Or should one condemn an economic system that gave some

families mansions on Summit Avenue and left others so poor that they would risk their children and their own lives for the sake of a single cow? They called it "The School Children's Blizzard" because so many of the victims were so young—but in a way the entire pioneer period was a kind of children's disaster. Children were the unpaid workforce of the prairie, the hands that did the work no one else had time for or stomach for. The outpouring of grief after scores of children were found frozen to death among the cattle on Friday, January 13, was at least in part an expression of remorse for what children were subjected to every day—remorse for the fact that most children had no childhood. This was a society that could not afford to sentimentalize its living and working children. Only in death or on the verge of death were their young granted the heroine funds, the long columns of sobbing verse, the stately granite monuments. A safe and carefree childhood was a luxury the pioneer prairie could not afford.

❦ ❦

"The dark, blinding, roaring storm once experienced, ever remains an actual living presence, that has marked its pathway with ruin, desolation and death," wrote South Dakota historian Caleb Holt Ellis in 1909. "The 12th of January, 1888, is, and long will be, remembered, not only by Dakotans, but by many in the northwest, not for the things we enjoy, love, and would see repeated; but for its darkness, desolation, ruin and death, spread broadcast; for the sorrow, sadness and heartache that followed in its train." To this day, nearly a century after Ellis wrote these words, the storm remains "an actual living presence" in the region. Mention the date to anyone whose family experienced the storm and you'll get a story of death or narrow escape. "There are those who say that that storm was no worse than others we have had," wrote Austen Rollag fifty years later, "but those who speak thus could not have been out of the house but sitting around the stove. I have seen many snowstorms in the more than sixty years I have been living here, but not one can compare with the storm of January 12, 1888."

The memories still burn. They burn all the fiercer because sorrow, sadness, and heartache did indeed follow in the blizzard's train. Drought ravaged the prairie in the early 1890s. Thousands who had borrowed against their homesteads went bankrupt in the financial panic that inaugurated the depression of 1893. Farm income slipped steadily in the last decades of the nineteenth century. The price of corn fell by half between the mid-1870s and the 1890s. A great exodus commenced on the prairie. By the time the rains returned late in the 1890s, over 60 percent of the pioneer families had abandoned their homesteads. Settlers came back, tried to make a go of it in the Dakotas or even farther west—and once again got burned out, frozen out, and blown away. Out-migration is on the rise once more. Nearly 70 percent of the counties in the Great Plains states have fewer people now than they did in 1950. These days nearly one million acres of the plains are so sparsely populated that they meet the condition of frontier as defined by the Census Bureau in the nineteenth century. Seven of our nation's twelve poorest counties are in Nebraska. As whites flee to cities and coasts, Native Americans and the bison that sustained them for thousands of years are returning. Indian and buffalo populations have now reached levels that the region has not seen since the 1870s. The white farmers and townspeople who remain would shun you for daring to say it, but in large stretches of prairie it's beginning to look like European agricultural settlement is a completed chapter of history. "It's time for us to acknowledge one of America's greatest mistakes," wrote Nicholas D. Kristof on the op-ed page of the New York Times, "a 140-year-old scheme that has failed at a cost of trillions of dollars, countless lives and immeasurable heartbreak: the settlement of the Great Plains."

The blizzard of January 12, 1888, was an early sign of that mistake. In the storm that came without warning, the pioneers learned that the land they had desired so fervently and had traveled so far to claim wasn't free after all. Who could have predicted that the bill would arrive with a sudden shift of wind in the middle of a mild January morning? A thousand storms of dust and ice and poverty and despair have come and gone since then, but this is the one they remember. After that day, the sky never looked the same.

Sources

I drew on a wide range of sources in writing this book—interviews with descendants of people caught in the storm, firsthand accounts archived at state and regional historical societies and research centers, contemporary newspapers, as well as secondary sources.

The archives I found richest in information about the storm were the Minnesota History Center in Saint Paul; the Norwegian American History Association in Northfield, Minnesota, the Minnesota State Climatology Office at the University of Minnesota, Saint Paul, Division of Library/Archives of the Nebraska State Historical Society; Holt County Historical Society in O'Neill, Nebraska; the Seward County Genealogical Society in Seward, Nebraska; the South Dakota State Historical Society in Pierre, South Dakota; the Center for Western Studies at Augustana College in Sioux Falls, South Dakota; the Dacotah Prairie Museum in Aberdeen, South Dakota; the Heritage Hall Museum and Archives in Freeman, South Dakota; the Dakotaland Museum in Huron, South Dakota; and the Jerauld County Historical Society in Wessington Springs, South Dakota.

The National Archives and Records Administration (NARA) in College Park, Maryland, holds the records of the Weather Bureau (Record Group 27) that I sifted through in order to piece together the story of how the storm was forecast and by whom.

In the notes that follow, I list and discuss the most important source materials I used for each chapter. I footnote only lengthy quotations, controversial issues, incidents or events for which I have found conflicting accounts or claims, and prickly subjects that have stirred up debate or confusion among historians of the period.

The only other adult book devoted entirely to this storm is *In All Its Fury: A History of the Blizzard of January 12, 1888,* collected and compiled by W. H. O'Gara (Lincoln, Nebraska: Union College Press, 1947), which I found invaluable for its account of the evolution and track of the storm, the historical context, and the extent of its impact.

Prologue

Books consulted for historical background on the Gilded Age include *America in the Gilded Age: From the Death of Lincoln to the Rise of Theodore Roosevelt* by Sean Dennis Cashman (New York: New York University Press, 1993); *America's Gilded Age: An Eyewitness History* by Judith Freeman Clark (New York: Facts on File, 1992); *The Gilded Age,* H. Wayne Morgan, editor (Syracuse, N.Y.: Syracuse University Press, 1970); *Unity and Culture: The United States, 1877–1900* by H. Wayne Morgan (Harmondsworth, England: Penguin Books, 1971), and *The Gilded Age: Essays on the Origins of Modern America* edited by Charles W. Calhoun (Wilmington, Del.: Scholarly Resources, 1996).

The claim that forecasts were correct 83.7 percent of the time comes from *Annual Report of the Chief Signal Officer of the Army to the Secretary of War for the Year 1888* (Washington, D.C.: U.S. Government Printing Office, 1888), pages 9–10.

Quote from Major John Wesley Powell beginning "[T]he promise of a science of profound interest . . ." comes from *The Chautauquan,* "National Agencies for Scientific Research," December, 1891, pages 291–97.

CHAPTER ONE

Departures and Arrivals

For the story of the Rollag family, I drew on family papers collected and translated from the Norwegian by Brynhild Rowberg of Northfield, Minnesota. Additional information and details about Norwegian immigrants come from *No Drum Before Him* by June Tisland (Madison, S. Dak. : Hunter Pub. Co., 1983); *Grass of the Earth: Immigrant Life in the Dakota Country* by Aagot Raaen (New York: Arno Press, 1979), and the memoirs and letters of Lars A. Stavig held at the South Dakota State Historical Society and at the Center for Western Studies at Augustana College in Sioux Falls, South Dakota. The detail about young Norwegian men building and storing their coffins comes from Stavig's papers.

My account of the history of the Swiss-German Mennonites, their migration to America, and their experiences in the communities around Freeman and Marion, South Dakota, relied on many conversations with and countless e-mails from Duane Schrag of Freeman, South Dakota. I also interviewed Gladys Waltner and her sister Anna Kaufman, nieces of three of the boys caught in the storm. I relied on the following books for additional details: *After Fifty Years* by John J. Gering (Marion, S. Dak.: Pine Hill Printery, 1924); *Memoirs of Rev. John Schrag and Family* (privately printed); *Our People and Their History* by P. R. Kaufman, translated by Reuben Peterson (Sioux Falls,

S. Dak.: Augustana College Press, 1979); *Looking Back 100 Years 1880–1980, Salem-Zion Mennonite Church of Freeman, South Dakota,* compiled and edited by Nita M. Engbrecht (Freeman, S. Dak.: Pine Hill Press, 1980); *The European History (1525–1874) of the Swiss Mennonites from Volhynia* by Martin H. Schrag, edited by Harley J. Stucky (North Newton, Kans.: Swiss Mennonite Cultural & Historical Association, 1974). Martin H. Schrag's book, pages 129–30, supplied several details about the journey from the Ukraine.

Details about the conditions on board ships of the Inman line are from *Travelling by Sea in the Nineteenth Century: Interior Design in Victorian Passenger Ships* by Basil Greenhill and Ann Giffard (New York: Hastings House, 1974), page 52.

Background on Lena Schlesselmer, later known as Lena Woebbecke, came from an interview with Lawrence Woebbecke at his farm outside Seward, Nebraska. I found additional information with the help of Jane Graff at the archives of the Seward County Genealogical Society.

Material on the Allen family comes from an interview with Barbara Allen Wegner at her home in Groton, South Dakota, and from copies of unpublished memoirs and recollections written by her father, Walter J. Allen.

Background on Benjamin Shattuck's Civil War service with the Seventy-third Ohio and his imprisonment on Belle Isle comes from Whitelaw Reid's *Ohio in the War* (Columbus, Ohio: Eclectic Publishing Co., 1893), Samuel H. Hurst's *Journal-History of the Seventy-Third Ohio Volunteer Infantry* (Chillicothe, Ohio: 1866), Warren Lee Goss's *The Soldier's Story of His Captivity at Andersonville, Belle Isle, and Other Rebel Prisons* (Boston: I. N. Richardson, 1874), and John L. Ransom's *Andersonville Diary* (New York: Haskell House, 1974).

Quote from railroad pamphlet beginning "Indeed, it may be justly claimed . . ." is from *Northern Dakota: Its Soil, Climate and Productions* by Northern Pacific Railroad Co., printed in Fargo, Dakota Territory, in 1877.

CHAPTER TWO

Trials

Details on grasshoppers are from *The Last Prairie: A Sandhills Journal* by Stephen R. Jones (Camden, Maine: Ragged Mountain Press/McGraw-Hill, 2000) and *After Fifty Years* by John J. Gering. I also found many pioneer accounts of grasshoppers and prairie fires in the archives of the Minnesota History Center and the South Dakota State Historical Society.

For the description of the sight and sound of a blizzard I drew on the story "Genesis" by Wallace Stegner in *Collected Stories of Wallace Stegner* (New York: Penguin, 1991), page 424.

Details about the blizzard of 1873 in Minnesota come from original accounts at the Minnesota History Center as well as Gilbert C. Fite's *The Farmers' Frontier: 1865–1900* (Holt, Rinehart & Winston, 1966).

The quote from *The Long Winter* by Laura Ingalls Wilder (New York: Harper & Brothers, 1940) appears on pages 134–35.

For my description of the Winter of Blue Snow I draw on material in Edmund Morris's *The Rise of Theodore Roosevelt* (New York: Coward, McCann & Geoghegan, 1979); "The Hard Winter and the Range Cattle Business" by Ray H. Mattison in *The Montana Magazine of History* 1 (October 1951), pages 5–21; Dee Alexander Brown's *Trail Driving Days: The Golden Days of the Old Trail Driving Cattlemen* (New York: Scribner's, 1952); and Ian Frazier's *The Great Plains* (New York: Farrar, Straus & Giroux, 1989).

Quote about the morning of January 12, 1888, beginning "On the morning of Thursday . . ." is from N. J. Dunham, *A History of Jerauld County, South Dakota* (Wessington Springs, S. Dak.: 1910), pages 164–5.

CHAPTER THREE

Disturbance

The description of the evolution and movement of the storm is based on interviews with Dr. Louis Uccellini, director of the National Centers for Environmental Prediction, Dr. Nicholas A. Bond of the Joint Institute for the Study of the Atmosphere and Ocean of the University of Washington and the Pacific Marine Environmental Laboratory of NOAA, Dr. Mark Seeley of the University of Minnesota, and Dr. Gregory J. Hakim of the University of Washington Department of Atmospheric Sciences. Greg Spodan and Peter Boulay of the Minnesota State Climatology Office provided useful background, statistics, and insight, as did Tom St. Martin. I also consulted *Meteorology Today: An Introduction to Weather, Climate, and the Environment* by C. Donald Ahrens (Minneapolis/St. Paul: West Publishing, 1994).

I relied on my interview with Dr. Uccellini for the description of the role of the jet streak and short wave in amplifying the storm. Dr. Uccellini pointed out that not all meteorologists share his views of this feature. Some prefer to look at the jet streak and short wave as a coupled phenomenon: Operating in tandem, the jet deepens the valley or trough of the short wave while the short wave steers the jet.

CHAPTER FOUR

Indications

Storm Watchers: The Turbulent History of Weather Prediction from Franklin's Kite to El Niño by John D. Cox (Hoboken, N.J.: John Wiley & Sons, 2002), *Isaac's Storm* by Erik Larson (New York: Crown, 1999), and *Air Apparent* by Mark Monmonier (Chicago:

University of Chicago Press, 1999) all provided a wealth of information on the history of weather forecasting and the ups and downs of the Signal Corps.

The bulk of my information on the career of Thomas Mayhew Woodruff comes from NARA, Record Group 27. Tom St. Martin has assembled extremely useful unpublished materials on the Minnesota State Weather Service, the career of Professor William Payne, and his stormy relationship with Western Union and the Signal Corps. The NARA holdings also shed light on the career and character of Adolphus W. Greely, as do *Ghosts of Cape Sabine* by Leonard F. Guttridge (New York: Berkley Books, 2000) and *A History of the United States Signal Corps* by the editors of Army Times (New York: Putnam, 1961).

Greely's assertion of "The great advantages of knowing sixteen to twenty-four hours in advance . . ." is from *Annual Report of the Chief Signal Officer,* cited above, page 12.

Chief Joseph's speech, "I am tired of talk that comes to nothing . . . ," is quoted in compiler Virginia Irving Armstrong's *I Have Spoken: American History Through the Voices of the Indians* (Chicago: Sage Books, 1971), page 116.

Woodruff's remarks about the Indian wars, "These wars are not welcome . . . ," are from his essay "Our Indian Question" in *Journal of the Military Service Institution of the United States* 2, no. 7 (1881), page 301. The second quote, "Under the banners of civilization . . ." is from page 295.

As background to Payne's objection to Western Union, it's helpful to know that by the terms of the Telegraph Act of 1866, private telegraph companies like Western Union charged the U.S. government a special rate fixed yearly by the postmaster general in exchange for the privilege of constructing and operating telegraph lines on public land.

In the matter of Woodruff's refusal to expand the data-gathering network by opening additional stations, what's especially puzzling is that Greely himself informed Woodruff on November 23, 1887, that he did "not coincide in the opinion expressed by you that the collecting of reports from the stations on the different railways free of expense to this service is contrary to the agreement with the WUTC [Western Union Telegraph Company] by the railway companies in question. Any business which is railroad business (and none other) can be transacted, and it is to be expected that you will avail yourself of such facilities as the railway companies offer in shape of free reports or information." A bit of background culled from the book *Old Wires and New Waves: The History of the Telegraph, Telephone, and Wireless* by Alvin F. Harlow (New York: D. Appleton-Century Co., 1936), pages 213–14, clarifies some of this. After the Civil War, the railroads and the telegraph companies had worked out a reciprocal arrangement whereby they both strung lines along the railroads' rights of way and used the same telegraph operator. The telegraph companies furnished the poles, wires, and instruments for both sets of lines. In exchange, the railroads agreed not to send any free messages except those pertaining to their own business. So, in effect, Greely was agreeing with Payne that it was permissible for Woodruff to receive weather data over the railroad lines for free so long as such data were used *only* in business pertaining to "the interests of the railways themselves" and *not* made available for public distribution. As a sop to civic responsibility, Greely added rather lamely that the public would stand to benefit "incidentally" from this arrangement by "the increased accuracy which such reports will give you in your work." Yet despite Greely's new instructions, Woodruff held firm in his refusal to avail himself of these data. Rather than run the risk of incurring a charge against the government, the lieutenant stuck by his original orders and limited himself to reports from the observing stations already in place.

Elias Loomis's observations that "[O]ne storm begets its successor . . ." is quoted in *Storm Watchers* by Cox, page 48.

Greely's remarks on the paths of cold waves are from his *Ameri-*

can Weather (New York: Dodd, Mead & Co., 1888), page 215. His quote about "principles of philosophy" is from *American Weather*, page 1.

Nicholas Bond explained to me that the spike in temperature at Helena during the first hours of January 12 was very likely the result of downslope winds blowing off the front range of the Rockies just west of the city, though Woodruff had no way of knowing this.

CHAPTER FIVE

Cold Front

The comparison of fronts to "seams in the atmosphere" and other details about the appearance and behavior of fronts come from John D. Cox's *Storm Watchers,* pages 165–66.

CHAPTER SIX

Explosion

In assigning blame for the lateness of the arrival of the cold wave warnings, it is critical to know exactly what time the messages went out and who caused the delay—but this is impossible to ascertain from the existing records. After comparing contradictory or obscure reports I have drawn conclusions that I believe are warranted, but in fairness I want to present all the evidence.

It's notable that in his report on the Minnesota State Weather Service printed in the *Annual Report of the Chief Signal Officer of the Army to the Secretary of War for the Year 1888,* Professor Payne wrote that after Western Union took over telegraphic work from the railroads on September 25, 1887, "The following circular was sent to all observers in Iowa, Wisconsin, Minnesota, and Dakota . . . : 'Messages will be ready for delivery at 7.30 a. m. daily (except Sunday) and should be called for before 8 o'clock a. m., which is the time for display of flags.' Very unsatisfactory service from that day to the present time has been given by the Western Union Company" (page 108). Payne is referring to messages sent to the "flag stations" of the Minnesota State Weather Service, but presumably the Saint Paul Indications Office would have followed a similar schedule.

As for the timing of the cold wave warnings sent out on January 12, the records that survive are contradictory. According to the monthly report that Woodruff sent to Greely for January 1888, the warnings to hoist cold wave flags on January 12 went out at 7 A.M. to Saint Vincent, Fort Totten, Fort Buford, Bismarck, Moorhead, Rapid City, Fort Sully, Huron, Valentine, Yankton, North Platte, Omaha, and Crete. And at 5 P.M. that day another round of cold wave orders was sent to Duluth, Saint Paul, LaCrosse, Dubuque-Des Moines, Davenport, Keokuk, Green Bay, and Milwaukee. Yet no observer in Nebraska or Dakota Territory received the reports before 12:20 P.M. Central time. By his own account, Woodruff did not arrive in his office until 9 A.M. and the records clearly indicate that he had not ordered the cold wave warning the night before. He was the only person in a position to authorize a cold wave warning, but he was not even in the office at the time that the first round of cold wave warnings were supposed to have been sent out. So how could the warnings have gone out at 7 A.M.?

It's notable that no other cold wave warnings that month went out before 11 A.M. In the tissue paper copies of Woodruff's indications, the first use of the term "cold wave" in connection with the storm was clearly dated and timed January 12, 10:30 A.M. (Central time)—which is consistent with an 11 A.M. transmission and a receipt shortly after noon. It's also worth noting that the entry in the monthly report to Greely detailing the transmission of cold wave warnings for January 12 is not in Woodruff's handwriting.

In sifting through and comparing the documents, I have concluded that the claim of the 7 A.M. transmission was incorrect—either a clerical error in the monthly report or, though this is far-fetched, a deliberate attempt to fabricate an earlier time and thus exonerate the Saint Paul office from responsibility for the deaths caused by the blizzard. In any case, despite this claim in the monthly report, there is no evidence that the warnings went out that early. And even if they did, the entries in the daily journals of Signal Corps observers indicate that the warnings were received far too late to benefit the people of the region, particularly the children who had left for school hours before.

Sergeant Samuel W. Glenn's fascinating and detailed firsthand account of the storm is in the Huron Signal Corps station journal for 1888 held at NARA.

"We were all out playing in our shirt sleeves . . ." O. W. Coursey, quoted in *In All Its Fury,* page 38.

Walter Allen wrote an account of how he got lost in the storm that day. Allen's daughter Barbara Wegner very kindly let me have a copy of the typescript. The Dacotah Prairie Museum also holds a transcript of an interview with Walter Allen about his experience in the blizzard, conducted on February 10, 1970, by Helen G. Strauss as part of the South Dakota Oral History Project.

There are several accounts of how the five Schweizer boys got lost in the storm. I relied heavily on the account written by Peter Albrecht, the brother of one of the storm victims. Additional information comes from the histories of the Schweizers in Freeman, South Dakota, cited above, and from my interviews with Duane Schrag, Gladys Waltner, and Anna Kaufman. Duane Schrag drove and walked with me over the route between the schoolhouse and the place where the boys were finally found. He also showed me the grave where they were buried.

I have used a bit of poetic license in imagining what was going through the mind of Johann Albrecht as he walked to school that morning and exactly how it came to pass that five of the boys split off from their teacher and two classmates—but in imagining these scenes I have been guided by suggestions in firsthand accounts of survivors and by my interviews with family and community members. Again, Duane Schrag was invaluable in verifying details about how these boys would have spoken, dressed, eaten, behaved, regarded their teacher, prayed, etc.

The story of Etta Shattuck comes from contemporary newspapers and bits of family history archived at the Holt County Historical Society and the Seward County Genealogical Society. I also drew on the accounts in *Homestead Fever* by Marie Kramer (Henderson, Nebraska: Service Press, 1993) and *Before Today: A History of Holt County Nebraska* by Nellie Snyder Yost (O'Neill, Nebr.: Miles Publishing Co., 1976). John Gilg drove me through the Holt County countryside where Etta Shattuck's school was located and supplied many useful details. Jane Graff showed me Etta Shattuck's grave in Seward and answered questions about her family's life in town.

Heloise Bresley of Ord, Nebraska, supplied the various—and markedly different—accounts of Minnie Freeman's rescue of her pupils during the storm. These accounts are held at the Ord Township Library. Depending on which source one consults—*The Trail of the Loup; being a history of the Loup River region* by H. W. Foght (Ord. Nebr: 1906), *A View of the Valley* compiled by the Centennial Committee in 1972, or *Scratchtown: A History of Ord, Nebraska* by Ronald J. Radil (Ord: Quiz Graphic Arts, 1982)—there were anywhere from nine to sixteen students present that day, and they either were or were not roped together by Minnie Freeman. Even the students present that day left contradictory accounts.

The story of Lena Woebbecke on the day of the storm comes from contemporary newspapers, the Seward County Genealogical Society archives, and the recollections of Lawrence Woebbecke. A Nebraska student named Hervey S. Robinson wrote a prize-winning essay in 1940 about Lena titled "A School Child in the Blizzard of '88," held at the Nebraska State Archives.

Arlein Fransen of the Jerauld County Pioneer Museum in Wessington Springs, South Dakota, provided newspaper accounts detailing the fate of May Hunt and her students in the Knieriem School during the storm. Duke Wenzel devoted much of the November 26, 2002, edition of the *Wessington Springs True Dakotan* to the storm and the Knieriem School, with photos of the site of the school and the surrounding countryside as they look today. I found this extremely useful. There is an extensive and detailed chapter about the storm in *A History of Jerauld County* by N. J. Dunham (Wessington Springs, S. Dak.: 1910).

CHAPTER SEVEN

God's Burning Finger

Professor John Hallett of the Desert Research Institute, Roger Reinking, and Ronald L. Holle explained the causes of the static discharges that so many people noticed at the height of the storm.

The quote from Herman Melville's *Moby-Dick* is from the Library of Literature edition (Indianapolis: Bobbs-Merrill Co., 1964), page 639.

CHAPTER EIGHT

Exposure

In describing how hypothermia affects and finally kills the human body, I drew on the following sources: *To Build a Fire and Other Stories* by Jack London, especially the stories "The Law of Life" and "To Build a Fire" (New York: Bantam, 1986, reprinted from *Novels and Stories by Jack London,* Library of America, 1982). *Hypothermia, Frostbite and Other Cold Injuries,* by James A. Wilkerson, M.D., Cameron C. Bangs, M.D., John S. Hayward, Ph.D. (Seattle, Wash.: The

Mountaineers, 1986); *Last Breath: Cautionary Tales from the Limits of Human Endurance* by Peter Stark (New York: Ballantine, 2001); *Hypothermia: Death by Exposure* by William W. Forgey (Merrillville, Ind.: ICS Books, 1985); *Hypothermia and Cold Stress* by Evan L. Lloyd (Rockville, Md.: Aspen Systems Corp., 1986); *High Altitude Medicine and Physiology* by Michael P. Ward, James S. Milledge, and John B. West, 3rd edition (London: Arnold, 2000).

I gleaned many useful details and corrected many mistaken ideas in the course of interviews with Dr. Cameron C. Bangs, Dr. Bruce Paton, Dr. Murray Hamlet, Dr. William W. Forgey, and Dr. Leona Laskin. In addition, Dr. Paton read the chapter and caught and corrected errors that slipped in.

Inevitably some of what I have written about the last hours of the Schweizer boys is based on speculation—but in all cases my speculations are based on firsthand accounts by hypothermia victims as well as the research and personal experiences of the doctors I interviewed. In several instances the doctors suggested the likely behaviors, conversation, and emotions of the boys during the various stages of their ordeal.

The statistic on bodily heat loss increasing as the square of the wind's velocity comes from *Hypothermia, Frostbite and Other Cold Injuries,* page 13.

Dr. Bruce Paton pointed out that if the boys had been wearing good windproof clothing, windchill would have had much less effect on them.

For the onset of deep hypothermia after the cessation of shivering, see William W. Forgey's *The Basic Essentials of Hypothermia* (Merrillville, Ind.: ICS Books, 1991), 31.

Details on hallucinations during hypothermia are from Evan Lloyd, *Hypothermia and Cold Stress,* page 140. Details on the mind separating from the body during extreme hypothermia are from Jack London, page 178, and *Into Thin Air* by Jon Krakauer (New York: Anchor Books, 1997), page 252.

CHAPTER NINE

Prairie Dawn

Drs. Bangs, Paton, Hamlet, and Forgey explained the likely causes of the deaths of Omar Gibson, Jesse Beadel, and Frederick Milbier.

My discussion of frostbite draws on the books cited above, along with interviews with these same doctors.

The fact that rubbing frozen flesh with snow causes tissue damage comes from *Hypothermia, Frostbite and Other Cold Injuries,* page 90.

Information on the history of inhalable anesthetics comes from Dr. Maurice Albin, Dr. Stanley Feldman, and Dr. Leona Laskin.

John Jensen's letter about the death of his wife and child was reprinted in a special edition of *The Wessington Springs True Dakotan* devoted to the blizzard, published on January 12, 1988, page 18.

CHAPTER TEN

Sunday

Nicholas Bond described to me the progress of vigorous cold fronts down through Texas and Mexico and out into the Gulf of Tehuantepec.

"There are very marked topographical features . . ." is from Greely, *American Weather,* page 212.

My account of the rescue of Etta Shattuck comes from *Before Today: A History of Holt County, Nebraska,* page 211. There is a markedly different account in the *Nebraska State Journal* of January 21, 1888, published in Lincoln. That paper reported that Etta was found by a Mr. Adams. As he was out shoveling snow off his hay pile, Adams got hold of a shoe, whereupon he asked, "Ettie, are you in there?" I chose the rescue by Daniel Murphy reported in *Before*

Today since it is more fully detailed and because the details match up with other contemporary sources.

Joan Killingsworth of Scribner, Nebraska, supplied family papers, stories and genealogies that I drew on in telling the story of her aunts, Eda and Matilda Westphalen. Ms. Killingsworth drove with me to the sites of the girls' graves and helped me find the location of their home, their school, and the place where they died.

CHAPTER ELEVEN

Heroines

My account of how newspapers reported the storm and created the heroines draws on a chapter in *In All Its Fury* about the storm coverage in the Nebraska press.

I was astonished, while reading through newspapers at the Nebraska State Historical Society, to come upon this item from the *Omaha Republican:* "Falls City, Nebr., Jan. 17 [1888]: A certain Mr. H. became intoxicated during the recent blizzard and in attempting to reach home, three miles south of town, had his hands so badly frozen that amputation is necessary. His attorney has brought suit against the saloon keeper for $5000." I had always thought that lawsuits of this sort originated in our own inanely litigious age. Of course, it's possible that this was a sick joke, for the same paper also ran this surely fabricated report: "Brainerd, Minn., Jan. 16: Lumbermen from Little Falls confirm the rumored murder of an entire family. Henry Ostrum murdered his wife and seven children because he feared they would freeze to death."

CHAPTER TWELVE

Aftermath

The Minnie Freeman story told in Florida comes from *The Sod House* by Cass G. Barns (Lincoln: University of Nebraska Press, 1970), page 112.

Greely's account of the storm is from *American Weather,* pages 223–24.

My account of the New York blizzard of March 1888 draws on *Blizzard!* by Jim Murphy (New York: Scholastic Press, 2000) and on my own account in *Braving the Elements* (New York: Doubleday, 1996).

"The holiday spirit of eight years before . . ." is from Hamlin Garland, *A Son of the Middle Border* (Lincoln: University of Nebraska Press, 1979), page 398.

The statistics on farm income and price of corn are from *The Gilded Age* by H. Wayne Morgan (Syracuse, N.Y.: Syracuse University Press, 1970), page 151.

Statistics on states with the poorest counties and population loss since 1950 come from "Amid Dying Towns of Rural Plains, One Makes a Stand" by Timothy Egan, *The New York Times,* December 1, 2003, page A1.

"It's time for us to acknowledge . . ." is from "America's Failed Frontier" by Nicholas D. Kristof, *The New York Times,* September 3, 2002, page A23.

Acknowledgments

The blizzard of January 12, 1888, is very much a part of the living history of the upper Midwest. Stories of relatives who were caught out in the storm are still told in hundreds of families—indeed, blizzard of '88 stories are part of the lore that defines a Dakota or Nebraska family. I'm deeply grateful to the individuals who shared their family blizzard stories with me: Gladys Waltner, Anna Kaufman, Brynhild Rowberg, Joan Killingsworth, Rita Hajek, Steve and Dawn Kenzy, Ardell Lovejoy, Barbara Allen Wegner, Robert Wegner, Lawrence Woebbecke, Delores Osborne, John Weinberg, Maureen Vig, Max Robinson, Helen Marie Kann Roche, David Mayberry, Ernest Spaulding, Orra G. Procunier, Russel Graham, Jan Holland, Arlene Tiffany.

My thanks to John Woodruff for supplying information about his great uncle, Thomas Mayhew Woodruff, and for giving me copies of the family trees of the Woodruffs and the Mayhews.

I want to give special thanks to three people whose expertise, advice, and generosity were invaluable during the researching and writing of this book. Duane Schrag of Freeman, South Dakota, opened the archives of the Heritage Hall Museum to me, welcomed me into his home, and answered an endless stream of e-mail queries about Mennonite customs, beliefs, and practices, the meanings of German words, and the complicated histories and genealogies of the Schweizer families I was interested in. Dr. Nicholas A. Bond of the Joint Institute for the Study of the Atmosphere and Ocean of the University of Washington and the Pacific Marine Environmental Laboratory of NOAA has been patiently fielding my questions about the atmosphere for years; his answers are invariably precise, clear, concrete, and colorful—a writer's dream. I can't count the number of times I asked him for help while I was researching and writing this book. Dr. Louis Uccellini, director of the National Centers for Environmental Prediction, started my education in atmospheric science when I turned up perfectly ignorant in his office many years ago. I'm extremely grateful for the brilliant tutorials he has given me whenever I manage to buttonhole him. Dr. Uccellini literally wrote the book on snowstorms, and I count myself fortunate indeed that he took the time to explain to me how the blizzard of January 12 evolved and why it developed with such violence and suddenness. Both Dr. Bond and Dr. Uccellini read drafts of my sections about the storm, offered suggestions, and corrected mistakes. Any errors that remain are entirely my own.

For additional help with climate statistics, meteorology, and the basics of the storm I would like to thank Al Dutcher, Kenneth G. Hubbard, Dennis Todey, John Hallett, Bob Henson, Greg Spodan, Peter Boulay, Tom St. Martin, Mark Seeley, Bruce Watson, Gregory Hakim, Roger Reinking, Thomas Schlatter, Ronald L. Holle, and Paul Kocin of the Weather Channel.

Dr. Cameron C. Bangs, Dr. Bruce C. Paton, Dr. Murray Hamlet, Dr. William W. Forgey, and Dr. Leona Laskin answered my questions about hypothermia and frostbite. I thank them all. Dr. Paton very kindly read and corrected drafts. Dr. Stanley Feldman, Dr. Maurice S. Albin, and Dr. Leona Laskin furnished information on the history of inhalable anesthetics.

I was extremely fortunate to encounter so many generous, helpful, and knowledgeable people at the various archives and historical societies where I did research. In particular I would like to thank Loris Gregory at the Minnesota History Center, Harry Thompson at the Center for Western Studies, Arlein Fransen at the Jerauld County Pioneer Museum, Marjorie Ciarlante at the National Archives and Records Administration in College Park, Maryland, the staff of the Nebraska State Historical Society, Kim Holland at the Norwegian-American Historical Association, Ruby Johannsen of the Dakotaland Museum in Huron, Sherri Rawstern and Kelly Face at the Dacotah Prairie Museum in Aberdeen, South Dakota, Carol Keyes and Kathy Manoucheri at the Holt County Historical Society, Jane Graff of the Nebraska History Network, Heloise Bresley of the Valley County (Nebraska) Historical Society, Don Zimmer of the Pierce Historical Society (Nebraska), Douglas P. Sall at the Dakota Territorial Museum in Yankton, South Dakota, Laurie Langland at Layne Library, Dakota Wesleyan University in Mitchell, South Dakota, Dick and Della Meyers of the Seward County (Nebraska) Historical Society, the staff of the Alexander Mitchell Library in Aberdeen, South Dakota, and the staff of the library of the United States Military Academy at West Point.

Friends and relatives provided invaluable leads, shared contacts, and wisdom, and furnished terrific advice. Many, many thanks to Bob Armstrong, Erik Larson, Pat Dobel, Phil Patton, Carolyn Ballo and her parents, Pete and Mary Ellen Kehn, Johanna Warness, Nils Dragoy, and Ivan Doig. My brother, Bob Laskin, my father-in-law, Lawrence O'Neill, and my brother-in-law, Lawrence O'Neill Jr. answered my dumb questions on matters mathematical, nautical, and scientific. My mother-in-law, Kathleen O'Neill, stumbled upon— and shared—an article explaining the origin of the word *blizzard*.

My mother, Dr. Leona Laskin, helped me research medical issues. My father, Meyer Laskin, inspired my obsession with weather in the first place and kept my nose to the grindstone with his inquiries about my deadline. Donald and Mary Kelly graciously provided hospitality while I was in Saint Paul. Avice Meehan, Bob Armstrong, and Jim Witkin entertained me during my research trip to Washington, D.C.

Thanks also to those who answered questions, sent me articles, furnished statistics, drove with me through the scene of the storm, and offered advice: Marie Kramer, Lori Ann Lahlum, Suzanne Bunkers, Curt Nickisch, Elizabeth Hampsten, Doug Foxgrover, William Rorabaugh, Paula Nelson, John Gilg, Lila Niemann, Nona Wiese, Marian Cramer, and Jean Magnuson at the United Methodist Church in Seward, Nebraska.

I have the great good fortune of enjoying two exemplary libraries very close to home—Suzzallo and Allen Libraries at the University of Washington (may its stacks ever be open!) and the King County Library System. I'd like to thank the staff of the KCLS interlibrary loan department for securing books and microfilm from all over the country.

My agent, Jill Kneerim, has been a dream to work with from the very start—passionate about good ideas, ferocious at the bargaining table, brilliant at editing, acute in her judgment and advice. When I hit a wall in the early stages of drafting the book, Jill pointed out an elegant way around it and, with a gentle shove, sent me on my way. For this and so much more, my profound gratitude. It was truly my lucky day when I found her (with a little help from Deb Brody).

Many thanks to Melissa Parker for her tireless efforts in helping spread the word of this book's publication in South Dakota, Nebraska, Minnesota, and Iowa. And thanks to Mary Whisner for volunteering to proofread, and to Megan Ernst for creating (and re-creating) the elegant map.

Tim Duggan, my editor at HarperCollins, has been an absolute pleasure to work with all the way through. His enthusiasm for the project has made every aspect of the research and writing that much more rewarding. Thanks also to Tim's assistant, John Williams, to my publicist, Jennifer Swihart, to my excellent copy-editor, Judy Steer, and to all at Harper who have seen the book through.

I've already thanked my parents, Meyer and Leona Laskin, for their help and advice. I'd like to thank them again for their incredible generosity through all these years. Kind and liberal in every sense, they are a continuing inspiration.

And finally, my own home team—my wife, Kate O'Neill, and our three terrific daughters, Emily, Sarah, and Alice, to whom this book is dedicated with love. Though they don't share my mania for weather, my wife and daughters are indulgent, sympathetic, and occasionally willing to be captivated by the fantastic displays the atmosphere whips up. I'm grateful to Kate for encouraging me to follow my dream and write about what I love. We've braved the elements together for a long time now—and there's no one else I'd rather be with through sun and storm.

Index

About the author

About the book

Read on

Insights,
Interviews
& More...

My Life at a Glance

THOUGH I HAVE NEVER LIVED on the American prairie, I have always been drawn to the weather and the landscape of this awe-inspiring part of the country: the immense sky that opens out from horizon to horizon, the summer thunderheads shadowing the distance, the radical plunges and rises in temperature, the inescapable wind, and the sideways snow. *The Children's Blizzard* gave me the chance to write about all of these—along with two other aspects of the prairie region that have long fascinated me: its history and the resilience, courage, and generosity of those who have made their lives beneath its fickle skies.

I have always loved to read, and I think my desire to write came directly out of the books that captivated me as a child. *The Adventures of Huckleberry Finn, Giants in the Earth, The Call of the Wild,* Tolkien's Lord of the Rings trilogy, *The Incredible Journey, Old Yeller, The Way West, King of the Wind*—these were among the books that started me on the path to authorship, and several of them ended up being inspiration or background for *The Children's Blizzard.*

I grew up on Long Island, New York, a child of the 1950s baby boom. Though my surroundings were clipped, tame, and leafy, my head was always full of Western adventure, Mississippi River boat trips, wagon trains, and sled dogs. We sang "America the Beautiful" in elementary school, and I still remember the sense of wonder I felt when I first saw "purple mountain majesties" in Upstate New York. I had always assumed this was just something

> 66 Two other aspects of the prairie region have long fascinated me: its history and the resilience, courage, and generosity of those who have made their lives beneath its fickle skies. 99

they put in a kid's songs—like a red-nosed reindeer and twinkling little stars. It astounded me that even relatively low-lying mountains like the Catskills and Adirondacks really are purple when seen from afar. It wasn't until my teenage years that I saw "amber waves of grain" on a bus and camping tour of the national parks. The bus drove as fast as possible from New York through the Midwest to reach the Badlands of South Dakota, and then the "real" mountains of Wyoming and Utah. But those amber waves and blue skies made an impression that stayed with me.

I was extremely fortunate in my formal education. I went to Harvard as an undergraduate and majored in the history and literature of England and the United States, and then spent two years at Oxford University, working toward a master's in English literature. For six years my primary responsibility was basically to read the finest works written in the English language and to delve into the history surrounding the creation of those works. The fact that all this glorious reading did not have a simple, straightforward "practical" application did not dawn on me until I left the ivory tower in 1977 and began looking for work. One prospective employer, a paperback publisher in New York, told me at my job interview that my education prepared me for nothing aside from being a terrible snob, but he took a chance and hired me anyway. I stayed in book publishing less than two years. I was restless. I quickly realized I wanted to be on the other side—writing the books, not editing and marketing them. So I took the plunge and hung out my shingle as a freelance writer. This was in 1980—a very long time ago, it seems to me now. The shingle has been hanging ever since. ▶

Meet *David Laskin*

© 2004 Jacqueline M. Koch

DAVID LASKIN is the author of *Partisans: Marriage, Politics, and Betrayal among the New York Intellectuals* and *Braving the Elements: The Stormy History of American Weather*. His writing has appeared in the *New York Times,* the *Wall Street Journal,* and *Smithsonian*. He lives in Seattle, Washington.

My Life at a Glance *(continued)*

My first projects were primarily books about raising children (my wife and I have three teenaged daughters), but I've always loved to travel and soon I was writing travel articles for the *New York Times* and then a book called *Eastern Islands* about the islands of the East Coast from Maine to Florida. My next book, *A Common Life,* which grew directly out of my education, was a group portrait of the friendships between eight major American writers, spanning the generations from Nathaniel Hawthorne and Herman Melville to the twentieth-century poets Robert Lowell and Elizabeth Bishop.

Soon after moving from New York to Seattle in 1993, I abruptly changed direction and plunged into a sweeping cultural history of weather in the United States. *Braving the Elements: The Stormy History of American Weather,* published in 1996, was the result. In the course of this narrative, I covered how weather has shaped every aspect of our nation's culture and history from the emergence of our identity as an independent nation to our efforts to forecast our violent and changeable weather, from our religious beliefs concerning weather to the various ways that radio and television have turned weather into news, entertainment, human interest stories, and just plain silliness. In my next book I used the same basic approach to focus on weather in the Pacific Northwest. *Rains All the Time* is an intentionally ironic title, for as I quickly discovered, the myth that the skies are always gray and dripping over Seattle and Portland is just that—a myth.

I never really planned to specialize as a writer, but over time it became clear that my

> **❝** Weather has shaped every aspect of our nation's culture and history from the emergence of our identity as an independent nation to our efforts to forecast our violent and changeable weather. **❞**

work fell into two broad categories: weather and literary biography. My next book, *Partisans: Marriage, Politics, and Betrayal among the New York Intellectuals,* belongs in the second category. Like *A Common Life,* it is a group portrait of American writers, though this time I looked at the tight-knit circle of men and women who dominated New York (and by extension American) literary politics from the late 1930s to the mid-1960s. *Partisans* won the Washington State Book Award in 2001.

After the publication of that book I was itching to find another weather-related subject. This time I wanted to move from history to story: I wanted to write a compelling true narrative about the impact of extreme and deadly weather on the lives of ordinary people who became extraordinary in the course of their encounters with elements beyond their control. That, in brief, was how I came to write *The Children's Blizzard.* ᗡᗡ

> ❝ Over time it became clear that my work fell into two broad categories: weather and literary biography. ❞

Finding and Telling
the Story

I FIRST LEARNED of the "children's blizzard" while writing *Braving the Elements*. I was working on a chapter called "Weather in the West"—about the droughts and tornados, dust storms and incredible cold fronts that American settlers marveled at when they crossed the Mississippi and began to settle on the great open prairie that stretches to the Rocky Mountains. As I searched for contemporary descriptions of some of these meteorological catastrophes, I came upon many accounts of the schoolchildren's blizzard, often written by the children themselves. These searing descriptions of the sky literally exploding with snow lodged in my mind. The image of children as young as five spending that frigid night of January 12, 1888, outdoors with no shelter haunted my imagination. My account of the children's blizzard took up a page or two in *Braving the Elements*, but the storm and the vividness of the accounts stayed in my head.

Another source of inspiration was Laura Ingalls Wilder's classic *The Long Winter*. This account of the snowy winter of 1880–81— a winter that began with a blizzard on October 15 and ended with floods the following spring—is both harrowing and absolutely factual. Every detail that Wilder put in the book—from the near starvation that isolated families suffered to the drifts of snow that buried their homes to the feeble fires of twisted hay they had to rely on for warmth—rings absolutely true to the accounts of the pioneers. I read this book to my daughters when they were young and I was mesmerized. My book grew directly out of my reverence for Laura Ingalls Wilder and my fascination with the narratives of the pioneers.

As the book gelled in my mind, I realized that it had all the elements of the books I most love to read and write: history, weather, religion, science, heartbreaking stories of struggle against the elements, intense faith, and bitter disappointment. What made it especially compelling to me was the fact that people confronted this storm as families. The white settlers, most of them immigrants from Europe or East Coast cities, had only been in this region for five or ten years. They were just getting established, paying off debts, replacing sod huts with their first frame houses—and then the blizzard came and destroyed them. Family was all they had, and it was the desire to save family members or the agony of wondering what had happened to a missing loved one that made these stories so overwhelming.

On an early research trip to the Nebraska State Historical Society in Lincoln, I filled a notebook with dozens of these family stories—and on subsequent trips to the South Dakota State Archives in Pierre and the Minnesota History Center in Saint Paul, I found scores more stories written by blizzard survivors.

I realized that the most gripping narratives in my book would be the ones with the most vivid details from families. I therefore decided to contact descendants of people who were caught in the storm, and to ask them whether they had family stories, memoirs, letters, and diaries about the event. Through the Internet I found a way to take out classified ads in every newspaper in Nebraska and South Dakota, ▶

> " My book grew directly out of my reverence for Laura Ingalls Wilder and my fascination with the narratives of the pioneers. "

Finding and Telling the Story *(continued)*

the two states hit hardest by the storm. The responses were phenomenal. After chatting with family members by phone or exchanging e-mail, I planned a series of trips to meet with them in their homes. These meetings were the most enjoyable part of researching the book. I sat down and talked for hours with people whose parents, aunts, uncles, or grandparents had been caught in this storm. I drove with them down country roads between fields of corn or soybeans searching for long-vanished country school houses or old homesteads. I walked the ground that their relatives had walked over one hundred years ago as the northwest wind tore at their faces and sealed their eyelids shut with shattered snow crystals. I searched out the churchyards where these storm victims lie buried. It was these meetings that really turned *The Children's Blizzard* from straight history into an intensely dramatic narrative that many readers have compared to a novel.

My cell phone also came in handy in tracking people down while I was on the move. I often found myself prowling through tiny country cemeteries, searching out the graves of storm victims and the nearby graves of their descendants. Some of the headstones bore the name of a person with the date of death left open. That's when I reached for my phone. Literally standing by her future grave, I placed a call to June Woebbecke and learned that she had married into the family of one of the most celebrated storm victims, a little German girl abandoned by her parents and raised by distant relatives. I'd read contemporary newspaper accounts of Lena Woebbecke's night alone on the freezing

> 66 I planned a series of trips to meet with [descendants of the storm's victims and survivors]. These meetings were the most enjoyable part of researching the book. 99

prairie, her terrible suffering and the eventual amputation of her legs. But I never expected to discover that the fifth generation of the Woebbecke family was living on that same farm, just two miles from the schoolhouse that Lena had left on that fatal day. Thanks to my interview with Lawrence Woebbecke, I learned that Lena had not died shortly after the blizzard, as contemporary newspapers reported, but lived on to young womanhood. She died shortly after her marriage and was buried in her wedding dress. Mr. Woebbecke gave me directions to Lena's grave at a tiny crossroads named Ruby not far from the interstate. The Immanuel Lutheran Church where Lena and her husband, George Schopp, worshipped is gone, but the graveyard remains. Somebody is looking after it, for the grass was clipped and the headstones stood straight. Standing there for a few minutes on a warm November morning and paying my respects to this brave and unfortunate individual was something I'll never forget.

By the time I finished writing the book, Lena Woebbecke and the other storm victims whose brief lives I had tried to reconstruct were precious parts of my own life. Writing the scenes of their deaths—or miraculous rescues—made the awesome, unpredictable power of America's weather almost unbearably real to me. ❧

> 66 Writing the scenes of their deaths—or miraculous rescues—made the awesome, unpredictable power of America's weather almost unbearably real to me. 99

Living History
The Author Receives Some Astoundingly Deep-rooted E-mails from His Readers

I HAVE RECEIVED scores of messages from families affected by the storm. Every time I read or lecture about the book, whether in the region where the blizzard hit or farther a field, someone comes forward to share a story about their storm-linked ancestor. Each story rings true to a detail or an account in the book. Each one is precious.

I'd like to share two of the family stories that directly concern individuals I wrote about in the book. I was touched that these people took the time to write me at such length. And it was affirming to see that what I wrote corresponds so closely with the stories that have come down through their families.

On New Year's Day 2005, I received this e-mail message from a woman living in Kansas named Donna Zerger:

Dear Mr. Laskin,

I picked up your book in a Barnes & Noble bookstore on a display of new books in Wichita, Kansas, and started scanning through the first chapter or two. I suddenly recognized the names of my husband's great-grandparents, Johann and Anna Kaufman, and I was immediately riveted. The story you tell is one that has been passed down in my husband's family for one hundred and sixteen years. His grandmother, Anna Kaufman Zerger, was the youngest of their ten children. We have heard mostly through oral history, how seven of her brothers died—one in the old country, one on the ship while crossing to America, one in the years following their arrival in South Dakota, three in the terrible blizzard, and finally the last one in Kansas of appendicitis at the age of twenty or twenty-one after working hard all day in the field. . . . The tragedies that dogged the Kaufman family through many years brought the three remaining siblings very close. . . . One hundred and ten years after the major disaster of the blizzard, the descendants of those remaining three siblings were still providing support and care to each other in times of need and celebration. The suffering they endured left a mark on the family, which has extended to three generations and shaped their relationships and connectedness.

The story was not new to me even before my marriage into the Zerger family, because I too am a "descendant" of the great blizzard of 1888. Peter O. and Susanna Gering Graber are my great-grandparents. My grandfather, Christian C. Graber, turned ten about a week after the fateful blizzard and was next in age to Andrew, who you call Andreas in the book. Although he was old enough to go to school and had been attending that year, he stayed home the day of the blizzard, but years later, did not remember why. His brother Peter was one of the five who froze to death and their house, being closest to the school, was the shelter they were all attempting to reach. What struck me when reading your book, was how closely all ▶

> 66 'Although he was old enough to go to school, he stayed home the day of the blizzard, but years later, did not remember why.' 99

the details you describe resemble the oral history which has been handed down through the generations in my family also. It seems amazing that one hundred plus years after the fact, the details that were handed down through two families mostly by word of mouth and that you discovered through extensive research, are so consistent. . . . John, the older of the two Graber boys who made it to their house and survived, grew up to become a Kansas state legislator and businessman. The family has been prolific and successful. Our branch is long on farmers, teachers, engineers, and carpenters and, includes architects, computer specialists, and meteorologists. From our humble beginnings in a two-room house, we found America to truly be a land of opportunity.

I thank you for choosing to research and write this story so that both of our family histories will be preserved.

—Donna Graber Zerger

A few days later an e-mail message arrived from a cousin of Donna's. Evidently the far-flung Graber family has a very active network. Carolyn Graber Trout's message to me was long and vivid and heartbreaking. Here is part of it:

Dear Mr. Laskin,

My dad told the story of the blizzard to us so often that we felt as if we had been there struggling in the blinding snow. His father—my grandfather, Christian

> 66 'My dad told the story of the blizzard to us so often that we felt as if we had been there struggling in the blinding snow.' 99

Graber—was home the day his brother Peter froze along with the other four boys. My parents have the brief description grandpa wrote of that terrible day. I don't remember grandpa telling the story, but I do remember hearing it from my great-uncle, Andrew, who managed to make it home safely. I was probably about six or seven when I heard the story from him, but I can still remember how he looked when he talked about letting go his brother's hand to wipe the ice from his face. Great-uncle Andrew was holding his cane, the tip of it centered between his shoes as he sat in his chair. When he was telling us how he couldn't find his brother's hand again, he rapped his cane against the floorboards. Hard. I remember that it made me jump.

That story of the blizzard was a particularly gripping and awful part of the family saga, but the entire family history has always held me in thrall. Our great-grandparents and their families suffered such terrible hardship and risked so much, but their strength was their legacy to us—all of us, their descendants keep their memories alive.

Carolyn goes on to describe a recent trip she and her husband, Tom, made to Freeman, South Dakota, in search of the Graber family homestead.

It was just across the section and down a bit from the church. There was still a farm at the location where the map indicated that my great-grandparents had lived. Tom would just have slowed and nodded at ▶

'Our great-grandparents and their families suffered such terrible hardship and risked so much, but their strength was their legacy to us.'

the mailbox, but I was driving, so there was NO chance of such a glancing encounter with my history. I pulled into the farmyard, got my camera, knocked on the door. The farmer and his wife were home and, although quite surprised to find a descendant of the original owner of the property on their doorstep, very gracious. Leroy Epp and his wife, Janette. Lovely people.

Leroy took us on a walking tour of the farm. We stood on the site of the school house (still visible as a faint depression in the ground), and I posed beside the school's cistern while Tom took my picture. We walked to the other side of the farmyard, about a quarter mile from the school, where Leroy had allowed a shelterbelt of trees to grow up around the foundation stones of the old homestead. I felt as though my blood were carbonated. It was such a short distance from the school to the homestead, just a few minutes' stroll for us on that sunny September morning. The terror of that storm was so long ago, but there we were, walking in their footsteps, imagining the suffocating snow and the immense cold.

—Carolyn Graber Trout

I'll close with one last story. In a book full of heartbreak, the incidents I found most upsetting to write about concerned children who survived the night of bitter cold on the prairie only to drop dead the next morning after getting up and taking a few steps. One such unfortunate was a young boy named ▶

> 66 The incidents I found most upsetting to write about concerned children who survived the night of bitter cold on the prairie only to drop dead the next morning after getting up and taking a few steps. 99